RESEARCH METHODOLOGY

[For MBA, M. Phil., B.Ed., M.Ed. according to the latest UGC Syllabus, also for MCA, CA, CS]

AF529341

Dr. VIJAY UPAGADE
M.Com., M.A., Ph.D.
Ex-Director Dr. S. Radhakrishnan College of Business Management
P/16, M.I.D.C. (Butibori)
Nagpur

Dr. ARVIND SHENDE
M.Com., M.A., Ph.D.
M.Phil., B.Ed., GDC & A
Principal & HOD of Commerce
Kamla Nehru Mahavidhyalaya
Nagpur

S. CHAND
PUBLISHING

S Chand And Company Limited
(ISO 9001 Certified Company)

S Chand And Company Limited

(ISO 9001 Certified Company)

Head Office: Block B-1, House No. D-1, Ground Floor, Mohan Co-operative Industrial Estate, New Delhi – 110 044 | Phone: 011-66672000

Registered Office: A-27, 2nd Floor, Mohan Co-operative Industrial Estate, New Delhi – 110 044
Phone: 011-49731800

www.**schandpublishing.com**; e-mail: **info@schandpublishing.com**

Branches

Chennai	:	Ph: 23632120; chennai@schandpublishing.com
Guwahati	:	Ph: 2738811, 2735640; guwahati@schandpublishing.com
Hyderabad	:	Ph: 40186018; hyderabad@schandpublishing.com
Jalandhar	:	Ph: 4645630; jalandhar@schandpublishing.com
Kolkata	:	Ph: 23357458, 23353914; kolkata@schandpublishing.com
Lucknow	:	Ph: 4003633; lucknow@schandpublishing.com
Mumbai	:	Ph: 25000297; mumbai@schandpublishing.com
Patna	:	Ph: 2260011; patna@schandpublishing.com

First Edition 2010
Second Revised Edition 2012
Revised Edition 2013
Reprints 2014, 2017, 2018

Reprint 2020

ISBN : 978-81-219-3222-6 **Product Code:** H8REM60BMGT10ENAC13O

PRINTED IN INDIA

By Vikas Publishing House Private Limited, Plot 20/4, Site-IV, Industrial Area Sahibabad, Ghaziabad – 201 010 and Published by S Chand And Company Limited, A-27, 2nd Floor, Mohan Co-operative Industrial Estate, New Delhi – 110 044.

PREFACE

Research Methodology is day by day becoming more important not only in the area of study but also in all the area of administration and management. Now on account of global competition traditional administrative and managerial practices has undergone several changes and this has made research methodology more relevent than before. It has become the internal part of the analytical system. Activities in Business Organisation, Government Organised and other sector are becoming more research oriented.

No policy can be decided by any of the organisation without collecting relevant facts and analysing them. The fact is also proved by providing the budeget for research in the organisation and setting up of Research and Development wing in the organisation.

There are too many books on Research Methodology in the area of social sciences. This book is a moderate attempt by authors to make the entire topic from indentification of problems to analysing the problem as simple as possible. We have made modest contribution in this direction specially for students and teachers community.

We are thankful to the Management Team and the Editorial Department of S. Chand & Company Pvt. Ltd. for all help and support in the publication of this book. We express our thanks to NYSS Institute of Management & Research for providing all possible help for completing the task.

We are also thankful to Hon'ble Shri Dattaji Meghe Ex - M.P. and Chairman, NYSS, Nagpur and Dr. Smt. Suhasani Wanjari, Chairperson, Kamla Neharu Mahavidhyalaya, Nagpur for constant encouragement.

We have no doubt in our mind that this book will serve as basic study material for the study of Research Methodology.

Dr. Vijay Upagade
Dr. Arvind Shende

This Page has been intentionally left blank

CONTENTS

CHAPTER 1

INTRODUCTION TO RESEARCH METHODOLOGY

After going through this chapter, you will be able to know :

- Introduction
- Definition
- Characterstics of Research
- Objectives of Research
- Nature of Research
- Importance of Research
- Relevance of Research
- Restrictions in Research
- Clasification of Research
- Research Process
- Stages in Research Process
- Research Process flow chart
- Research Problem
- Identification of the problem
- Steps in selecting Research Problem

Research work is traditionally defined as gathering of data to answer the questions and finding solution to problems. The research work may be asked to provide answers to questions of theoretical interest to particular discipline. This may include such questions which have no interest to the society.

In this chapter the aim is to provide a brief overview of research methodology along with simple model of research methodology. We will try to find out the importance of research methodology in handling the data.

The research in management and social science are concerned with human beings, which is attach to its surroundings. The organizational behavior is an important part of the focus of research in management and social sciences and researcher must consider this part carefully to come to conclusions.

The concepts of validity and reliability provide the criteria by which we need to judge our choice of research methods. These criteria determine the credibility and academic value of research work.

The validity means success of a method in assessing what is to be assess. If method is valid than results can be taken as true. There are four main approaches for assessing validity, viz face validity, content validity, predictive validity and construct validity.

The reliability is a criterion that refers to the consistency of the data stemming from the use of a particular research method. A measure is reliable to the extent that the repeated application of it under the same conditions gives the same result. However, whatever research method is used, it is always possible to consider carefully features of research design that make for reliability.

Research is commonly know as search for knowledge. It can be defined as search for systematic knowledge. Research is scientific and systematic investigation in relation to specific aspect. It is a movement from known to unknown.

Research is considered as careful investigation or inquiry to find out new facts in any branch of knowledge. It is an original contribution to the existing stock of knowledge making for its advancement. Therefore research is systemized effort to acquire new knowledge.

Research should not be considered as academic activity only, it also applies to all the Respect of human activity. Research is basis for making process of decision effective and more meaningful. It includes the defining of research problem, formulation of hypothesis, collection of data, analyzing of data and arriving to conclusions. After the conclusions are drawn, it is also necessary to find out testing of the conclusions in relation to the hypothesis.

Now research has acquired so much of prominence in all the activities, that no development can be thought without proper application of research.

Research is now taken up by many professional It is a way of thinking. It critically examine the various aspect of the professional work. It is a habit of questing. what you do and systematic examination of observed information to find answers, with a view to make appropriate changes for effective use of the information.

Definition

Thinkers who have contributed in defining the research are given below:

The word research is composed with two syllabus re and search as defined re means again and search means examine closely and carefully. So research means it is systematic study and investigation in some field of knowledge, undertaken to established facts.

Grinnel further adds

'Research is a structured inquiry that utilizes acceptable scientific methodologies to solve problems and creates new knowledge that is generally acceptable.'

Lundberg

draws a parallel between the social research process, which is considered scientific and the process that we use in our daily lives.

Burns defines

Research as a systematic investigation to find answer to a problem.

According *to Kerlinger*

'Scientific research is a systematic, controlled empirical and critical investigation of propositions about the presumed relationships about various phenomena

***Bulmer* states**

'Nevertheless sociological research is primarily committed to establishing systematic, reliable and valid knowledge about the social world.'

On the basis of the definitions given above we can summarize that the research is an organized inquiry designed and carried out to provide information for solving problem. It is a careful inquiry to discover new information to expand the existing knowledge. It is an investigation, recording and analyzing evidences for on going knowledge. The main thrust of research is on going knowledge to solve the problem. Unless the problem is properly assessed no appropriate solution can be provided.

On account of fast industrialization in post-independence period in India business in general was facing various complex situations. This has resulted in requirement for more accurate managerial and administrative decisions based on facts. This has created need for business decisions to be taken on the basis of research findings.

Characteristics of research

From these definitions it is clear that research is a process for collecting, analyzing and interpreting information to answer questions. But to qualify as research, the process must have certain characteristics : it must, as far as possible, be controlled, rigorous, systematic, valid, verifiable, empirical and critical.

Let us briefly examine these Characteristics for proper understanding

1. Controlled

There are many factors that affect an outcome. In a study of cause and effect relationships it is important to be able to link the effect (s) with the cause (s) and vice versa. The concept of control implies that, in exploring causality in relation to two variables, you set up your study in a way that minimizes the effects of other factors affecting the relationship. This is possible in physical sciences, where as in social sciences such controls are impossible, therefore attempts are made to quantify their impact.

2. Rigorous

You must be scrupulous in ensuring that the procedures followed to find answers to questions are relevant; appropriate and justified. Again, the degree of rigour varies between the physical and the social sciences.

3. Systematic

This implies that the procedures adopted to undertake an investigation follow a certain logical sequence. The different steps cannot be taken in a haphazard way. Some procedures must follow others.

4. Valid and verifiable

This concept implies that whatever you conclude on the basis of your findings is correct and can be verified by you and others.

5. Empirical

This means that any conclusions drawn are based upon hard evidence gathered from information collected from real life experiences or observations.

6. Critical

critical scrutiny of the procedures used and the methods employed is crucial to a research inquiry. The process of investigation must be foolproof and free from drawbacks. The process adopted and the procedures used must be able to withstand critical scrutiny.

Objectives of Research

The research is being conducted with a purpose of discovering answer to the question by making an application of scientific procedures. The main object of the research work is to take out the hidden facts yet to be discovered. Every research study has its own purpose to be attended, therefore objectives of research can be broadly classified as academic and utility one.

A. Academic Objectives – It relates to development of new concept and addition to old concept. It means that urge for knowledge is main factor in this type of objectives of.research.

B. Utility Objectives – It relates to utility of research work, as research is accepted for more use to the society. It has to provide the base for policy formation in the society.

This type of research must be in a position to provide for achievements in attending organizational objectives.

As explained above the research provide base for investigation by which relationship between two variables can be established. It is only in systematic research where induction and deduction are possible.

Observation is main element in the research study by which decision making ability can be increased. Collection of primary and secondary data is involved in the research process and now with these requirements objectives of research can be further classified into following:

1. Decision Making Objectives
2. Environmental Objectives
3. Market Objectives
4. Customer objectives
5. Profit and Promotional Objectives

1. Decision Making Objectives

Decision making in now influenced by research. The project identification and implementation is based on the research conducted. There cannot be any business policy which is not affected by research findings. Controlling, which is the main function in the management, can effectively be organized through research study.

2. Environmental Objectives

All the decisions in the business are taken in relation to the environment in which business operates. All the factors affecting business like State, Investor, worker, customer and the competition requires systematic investigation before any decision is to be taken.

3. Market Objectives

The market objectives of research are defined as market research. This includes the market share of products, profit margin of the organization and total sales volume of the company. On the basis of the careful investigation of the available market information, relevant market strategies can be drawn regarding new product development, product selling approach and product modification.

4. Customer Objectives

The need of the customer is assessed, well in advance even before product is planned. The utilities of product is decided on the basis of the quality of the product, in relation to the requirements of the customers. It is in this respect that the inquiry is conducted to find out the level of satisfaction of customers.

5. Profit and Promotional Objectives

In most of the companies profit maximization is the main objective to be attended by them. This requires investigations and consultations to be conducted. Surveys are also conducted to work out the variables in support of the promotional activities. The research provide strong base for these activities. The development of business entity is based on corporate image which is outcome of the relationship between internal and external factors of the companies.

Nature of Research

Research is process in which in-depth study of the problem is carried out. This requires investigation to be conducted based on collection and compilation of data along with its interpretation and presentation.

Research, if properly conducted, is helpful in decision making process. There can not be any research exercise which will not yield any additional inputs to knowledge. As such research involves critical examination of facts which leads to formation of new concepts or modification of old concepts.

The research activities will help us in testing of hypothesis and establishing relationship between variables, by this we can identify the methods for solution of the research problems.

The research is a fact finding process, which influences the decision to be taken. This also provides an opportunity to check the effectiveness of the decision taken. The research is a scientific process and it is required to be conducted in proper sequences, which includes activities right from identification of research problem, formulation of hypothesis, testing of hypothesis, observation and relationship of variables and drawing of conclusions.

Importance of Research

All development and progress in the society is an out come of a research work. The development of logical thinking is promoted by conducting the research activities.

Research provides base for the policies formation of the Government related to agriculture, industries and infra-structural services in the region.

In this context investigation in the structure of economy is conducted through compilation of data and analysis of facts. Research provides for predicting of future prospects of the region.

Research has special importance in relation to solving various problems of business and industries. Market research,Operation research and Motivation research are conducted in the business for various requirements.

The importance of research can also be ascertain through its application in sociopolitical sector in order to find out solution to social and political problem of the society.

Relevance of Research

Though research is extensively used everywhere in business for planning, forecasting and decision making, but it may lose its relevance of business competition in the society.

Therefore in under developed countries normally having no competition it may considered as luxury and for this very reason it may be neglected in the society. In the existing stage of business competition, where problems are becoming more complex, research is becoming more relevant than before. Usefulness of research can very well be seen in the managerial functions.

Research establishes the relationship between variables and functional areas. It is an effective tool for forecasting. It also provides necessary help for the optimum utilization of the available resources. The formulation and implementation of policies and strategies will be more effective, when they are based on the research studies.

Research is main source of decision making as it helps the process of thinking, analyzing and interpretation of the business situations. It provides base for innovations product development and product modifications.

Restrictions in Research

The research suffered with certain restrictions in spite of its relevance and usefulness.The social research is subject to changes in the society. As the social structure changes, the decisions taken can not be implemented as it is. There are several other factors which influences the decisions. Unless these factors are properly analyzed decision may be biased.

Sometimes the authorities are also not serious about the implementation of the decision taken. If business activities are carried on the basis of custom and traditions than in such cases research study becomes irrelevant.

The research activities are very expensive, therefore it may not be viable for small and medium scale unit to take advantages of the facilities.

Classification of Research

The research activities are conducted with sole object of finding solutions to unsolved problems, so that the services to the society can be provided. There are various approaches for conducting research activities and these research can be divided into 3 different classifications. The classifications are shown below in the following figure:-

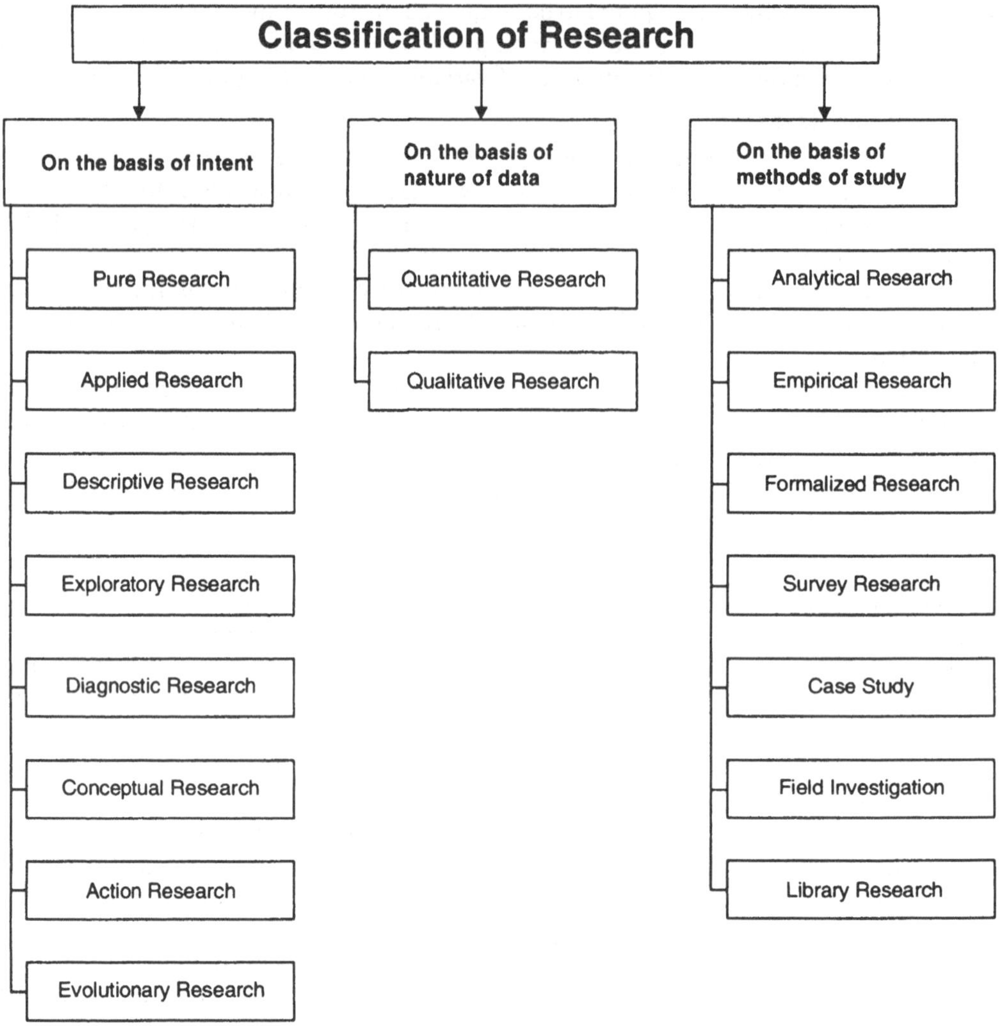

1. Pure and Applied Research

Pure research is also called as fundamental research. This is conducted with view for the sake of the knowledge having no intention of its application. This study is conducted for generalization which will help us to form certain theory. In pure research if study is conducted to assess human behavior, it may provide us generalization in relation to normal human behavior.

Pure research offer solutions to many practical problems. It helps us to find out various crucial factors. It develops many alternative solutions.

Applied research is carried out when any real- life problem or social problem requires immediate solution which can be helpful in policy formulation by the concerned agency. This is considered problem oriented and action directed activity in which immediate and practical results are required.

There is vast scope for applied research in the field of technology, management, commerce and economics, as it contributes to the development of new facts or testing of new theory.

Applied research can put theory to the test, it may add to conceptual clarification and integrate previous existing theories.

2. Descriptive and Analytical Research

Descriptive research is term as Ex-post- facto research. This type of research are mainly concerned with description of facts only. It is an expiation of state of affairs as they exists.This includes surveys and facts findings enquiries.

The main important aspect of this type of the study is that researcher has no control on the variables under investigation, he can only describe things as they exist and discover the causes.

Analytical research relates to critical evaluation of the existing facts and information which is used in order to draw certain conclusions.

3. Quantitative and Qualitative Research

Quantitative denotes methods that generate data comprising of numbers. This research is conducted for the measurement of quantity and it is applicable to the work which can be expressed in terms of quantity only.

Qualitative research relates to quality involved in assessment. It generates the data that is verbal. Assessment of human behavior is a qualitative research.

It is very difficult to find out which research is quantitative and which one is qualitative. There are four main methods for these types of research work namely – interview, questionnaires, observations and documents analysis. On the basis of their explanation none of this can be exclusively labeled as quantitative or qualitative. Accordingly, the labels of quantitative and qualitative should be applied to data rather than to methods.

Motive research is important type of qualitative research, which is conducted to assess the motive or desire of human behavior. Likewise Opinion research is also qualitative research conducted to know how people feel about particular subject.

4. Conceptual and Empirical Research

Conceptual research is conducted by the thinkers and philosophers for developing new theories or for reinterpreting old one.

Empirical research is based on observation or experience without due regards to theories. In this type of research working hypothesis is provided and then data is collected to prove or disprove this. Researcher here sets up the design for desired results.

5. Exploratory and Formalized Research

The research study in which hypothesis is developed is called exploratory research, where as when hypothesis is tested in the research study it is known as formalized research.

6. Survey Research

A survey is conducted with an object to understand specific aspect in defined population. Usually the population in study is so large that the access to all the elements is impossible.

Therefore sample taken from the population will make it possible to draw the conclusions regarding population. The purpose of the survey is to understand phenomena by identifying influencing factors.

Survey are normally conducted through questionnaires but some times interview method is used to collect necessary information. The method of conducting survey may also be applied to analyzing large data files in an organization by using sample form. Survey method is widely accepted for following reseaons.

1. Greater amount of objectivity.
2. Greater reality in findings and conclusions.
3. It is most scientific method for obtaining data and dependable result.
4. Direct observation of the population has possibility of getting accurate results.
5. Survey is extensively used in all the areas for its merit.

7. Case study

It is very difficult to define, when the phenomena under investigation are not readily form its context in such situation case study becomes appropriate. It is characterized by in-depth study of organization. It is a concept and intensive study of business situation.

Case study is method of investigation for exploring a live situation, it is a in-depth study of situation as whole. It is in the form of qualitative and quantitative analysis where careful observation of situation is done.

A case study uses various methods for collecting information may be in form of interview, questionnaires, observation and documentary analysis. By this, wide variety of relevant data is collected.

A case is report containing facts and opinion expressed by the people. The cases are designed to acclimatize the situation to participant. This provides the description of the decision taken in particular situation.

8. Action Research

When research is conducted for the study in a single workplace, than action research is commonly used in such cases. The focus in action research is given in improving working practices. The action research has practical purpose and it is directed towards organizational changes in managerial set-up.

The action research theories are valid through practice, rather than applied independently to practice. It is conducted through direct action. In action research, research is conducted in various phases like, base line survey, systematic action and periodical assessment.

9. Field Investigation

The field investigation is conducted in the assessment of actual situations. The results in this type of work is more realistic. In order to have these results, accuracy is required to be maintained by making selection of proper sample. In field investigation it is very difficult to control variables and informants in the field.

10. Evaluation Research

The evaluation of any project is the subject matter of the evaluation research. This type of study is conducted to evaluate ongoing projects or project already carried on. The performance evaluation, job evaluation and PERT (programme evaluation and review technique) are main aspects of the studies.

This includes periodical and terminal evaluation. It is very useful for the evaluation of actual operational efficiency.

11. Library Research

When secondary data is reliable enough to be used for drawing conclusions than library research is very useful. This type of research is based on the assessment of the data available from books, periodicals and journals available in the library.

The observations and conclusions drawn in these cases may be bias if data is not used carefully.

As this is only compilation of available data it is not considered as actual research work. If systematic investigation is made it can also be very useful research study.

RESEARCH PROCESS

There are various approaches to research work. Every research project is different from each other on account its time and place in which it is to be conducted. But all the research projects will have common understanding of the society and have common steps through which research is conducted.

In order to have proper research work, every research project have to have research problem identified. The research process will consist of particular objective which should be clearly stated and there should be a hypothesis which has to be proceed right or wrong. In addition to this each research project will have research design which indicates as to how required data will be collected, analyzed and interpreted.

The research process consists of series of various actions, which are necessary to effective research work. It includes all such stages required to carry out research work. This must specify desired steps involved in conducting research work.

It is necessary to understand that there is no specific sequence or established order in which research work is carried out. Therefore in research process there are certain guidelines regarding steps involved in research work.

The stages in research process are listed as, selecting research problems and stating of hypothesis, formulating of research design, collecting.analyzing and interpreting of data.

These steps can summarized as under

1. Formulate the research problem.
2. Get the background information.
3. Refine your search topic.
4. Consider your research option.
5. Select the appropriate tool (i.e. sampling procedure)
6. Use the tool effectively.
7. Locate your material (i.e. data for research)
8. Analyses the material located (i.e. data analysis)
9. Organise and write report.
10. References and bibliography.

The stages which are explained above are dependent on each other. It is clear from the above steps sequence that researcher who has no knowledge about subject matter will not be in a position to collect and analyze data.

Therefore researcher is suppose to have proper knowledge of later stages before be perform the earlier stages. Research process is system of interdependent related stages. Before giving brief description of the various stages involved in research process it is necessary to understand the difference between research method and research process.

Difference between Research Method and Research Process.

Research Method	Research process
1. Formulation, Analysis of information needed	1. Chose broad topic. Get overview of topic, narrow down the topic
2. Identification and Appraisal of the likely resources.	2. Formulate question to guide research plan for research
3. Tracing and locating individual resources.	3. Find analyze and evaluate the resources
4. Examiniing and selecting individual resources.	4. Evaluate, evidence take notes, compile the bibliography.
5. Recording and storing of information	5. Established conclusions and organized information.
6. Interpretation and analysis	6. Interpretation and analysis.
7. Shape Presentation and communication of information.	7. Creat and present final project
8. Evaluation of assignment	8. Reflection- satisfactory presentation.

Stages in Research Process

1. Chossing of Research Problem, Review of Literature and Hypothesis.
2. Research Design.
3. Sample Design and Collection of Data.
4. Analyzing of Data and Testing of Hypothesis.
5. Interpretation and Report writing

As explained above research process consists of various series of action, which are necessary to conduct research work effectively. Therefore research process includes all such steps necessary to carry out research activities. Research process also includes desired sequencing of all the steps involved in the research work.

Let this thing be very clear that there is no such specific sequence which is recommended any where in the research methodology regarding steps for the research process, where as there are certain guidelines, as to how to organize steps in the research process, which are described as under -

1. Formulation of research problem
2. Review of the existing literature
3. Formation and development of working hypothesis
4. Preparation of research design
5. Determining sample design
6. Data collection
7. Project execution

8. Data analysis
9. Testing of hypothesis
10. Data interpretation
11. Report of the research work

In order to understand the research process perfectly brief description of each of the stages will be of more help to researchers.

1. Formulation of the research problem

The research problem relates to statement of problem and relationship between two variables under study. Research has to identify the problem first and later on its required to single out the problem.

This will give scope to researcher to decide general area of interest or subject matter of the research study. It means researchers will have to narrow down scope of the study by formulating general topic of research study in to specific problem under study.

Therefore in relation to the formulation of research problem two steps are involved basic understanding of the research problem and rephrasing the research problem in to meaningful terms.

2. Review of the existing literature

Research can not be conducted without reviewing of existing literature which is available. The existing literature may be conceptual or empirical in nature. Conceptual literature is concerned with concept or theories. Empirical literature is concerned with earlier studies of similar nature already conduced. This review of existing studies provides base to understand how to plan for the study.

The review of the existing literature will help researcher to define his research problem of general nature into analytical terms or onto operational terms.

If the problem is properly defined on the basis of the review of the literature, this will help the researcher to discriminate the relevant data from collection of irrelevant data and confusion at this level can be avoided.

3. Formation and development of working hypothesis

After the formation of the research problem and reviewing of the available existing literature now hypothesis is required to be explained by the researcher. The normal assumptions which are the base of the study which may be tentative in nature are considered as hypothesis.

These assumptions are drawn to test its logical sequence.

The researcher should try to avoid vagueness in these assumptions. Therefore hypothesis should be as clear as possible, it should be very specific.

Hypothesis is a guiding force of the researchers, it is helpful in sharpening his thinking process and by this he will be in a position to focus his attention on the important facts of the research.

It is an outcome of researchers deep thinking about the subject matter of the research, that he is able to settle on formation and development of hypothesis.

4. Preparation of Research Design

The research design is prepared by the researcher after the formulation of research problem, reviewing of literature and developing of hypothesis. It is an outline or a conceptual structure within its limit research work is suppose to be carried on.

The research design is prepared with an object of collecting relevant data with the minimum efforts and with minimum of expenditure, just to control wasteful expenditure.

There are several methods for preparation of the research design. The important amongst them are experimental and non- experimental one. Experimental design can further be classified into formal or informal one.

It is left to the researcher to work out for selecting a proper design for research. Depending upon the utility and requirement a specific design should be selected.

The research design has to take in to consideration the purpose of the research study, which may be exploration, description, diagnosis and experimentation. The researcher has to make careful choice in selecting research design.

The research design requires certain facts to be considered like ability of the researcher along with his staff to be employed for research work. Time and cost available for research work is another factor to be considered. Data collection is the most important factor, in which methods for data collection and organization of collection is to be specified.

The research can be conducted for all the items under investigation which constitute universe. The census of population conducted by the Government of India is an example of such studies. This type of the study is very time consuming and has been very costly also. Therefore in place of the study of the whole universe sample study can be conducted for the purpose of the research.

5. Determining Sample Design

The success of the research study is largely based on proper identification of the sample to be selected for the study. The method for selecting is normally known as sample design. It is a sample plan already decided before data is collected from given population.

The samples can be classified as probability or non-probability one. In probability sample item of the population is likely to be selected through simple random sampling, systematic sampling, stratified sampling, cluster or area sampling. Where as non – probability sampling is conducted through convenience sampling, judgment sampling and quota sampling.

A brief description of these techniques of various sampling are given below-

A. Deliberate Sample

It is one of the non- probability sampling, also known as purposive sampling. This technique involved deliberate selection of sample which represent the whole universe. In case of data of LPG user, researcher may select few gas distribution centers to conduct interview of gas users. But in case of judgment sampling researcher's judgment is used for selecting sample.

B. Simple Random Sampling

This is one of the techniques of probability sampling, also known as chance sampling. As this sample is controlled by probability, there is every and equal chance for all the items of the universe to get selected.

C. Systematic Sampling

This is most simple and practical way of sampling. It is covered under the techniques of probability sampling, in which every 5th or 10th item of the universe is selected. If study of the sampling students of specific class is conducted then every 5th student on the roll is selected for the purpose of the study in systematic sampling. This study is possible only when source list for the sample is available.

D. Stratified Sampling

This is also a technique of probability sampling, which is used for the population not having homogeneous group. In this technique population is classified in small strata, which should not be overlapping and sample unit is selected from such strata. So in this process of sampling first stratification is made then simple random sampling is done.

E. Quota Sampling

This is an important form of non- probability technique of sampling. Quota sampling is considered as judgment sampling. If the study stratified sampling is becoming costly on account of sample from individual strata than the enumerators are given specific quota to be filled in from different strata. It is left to the judgment of the enumerators to decide on selecting the quota for the study. The size of the quota is based on the size of the strata.

F. Cluster Sampling

This is a probability sampling technique. Cluster refers to a group and in this sampling technique first the population is grouped, then specific group is selected for the study.

It is a group sampling rather than individual sampling and cluster is selected from random sampling. Suppose bank wants to sample its credit cards issued by them. It has issued cards to 30,000 customers, if sample size to be kept 900 cards so for the cluster sampling of 30,000 cards holder could be formed in to 100 cluster of 300 card holder each.

G. Area Sampling

The area sampling is just like a cluster sampling. When geographical area under the investigation is very large than the total area is divided in to non-overlapping small area. This division of total area in to a small one is called geographical cluster.

These small area so divided are selected randomly, in which efforts are made to select all the units in the sample.

H. Multistage Sampling

When the geographical area under investigation is very large, then development to area sampling is required. Therefore area sampling is developed to a multistage sampling.

In this sampling technique total geographical area is divided into small area. Such classification may be first region as State, which is further sub classified into district, then in taluka/ thehasil or further to a block of village lastly to group of families. In all the stages random sampling is applied in order to have multistage sampling.

I. Spot Sampling

When sample can not be decided at the time of the investigation and if it becomes very difficult to assess what type of information will be generated , in such complex situation no sampling technique is pre decided.

In spot sampling technique, sample design is decided as the survey progresses based on the information collected. The researcher must take into consideration all the factors like nature of inquiry and other related aspects to work out sample design.

6. Data Collection

The data is collected as per the requirement of the study. This may be primary or secondary in nature. If the secondary is sufficient enough to assess the research problem then there is no necessity for primary data to be collected.

When it is found that data in hand ie secondary data, is not sufficient to draw conclusions then there is no alternative except to for primary data to be collected by the researcher.

As per the requirement of the research study appropriate data is a basic need of researcher. This primary data can be collected by observation, interview, questionnaire and schedule. Each of this source of information of primary nature is describe in brief here.

A. By Observation

It is only by the observation of certain facts required information is collected. The observer is trained to record the necessary information, which will be used for analyzing the data.

The limitation of this method is that it can collect only current information. There is no provision as to how past or future information will be ascertained.

B. Through Interview

i. Through Personal Contact - This is second important method for collecting primary data.The investigator in this method will come in direct contact of the person from whom this information is to be collected.

This interview may be structured or unstructured one. In most of these cases the structured method is used for conducting interview. The investigator is suppose to collect information on the basis of the pre determine set of questions. Some times when enquiry is very simple then unstructured method is used, where the question are developed as the discussion progress. The utility of the information collected in this method is largely based on the skill of the investigator.

ii. Through Telephone - When the personal contact with person is not possible than in order to contact person telephone is normally used. Through this method maximum people can be contacted in very short time, but telephonic conversion has its own limitations. It is expensive and only few questions can be asked and it is very difficult to establish the identity of a person giving answers.

Now with the development of mobile and internet, information collected through this method has changed substantially.

C. By Questionnaires

When it is not possible for the researcher to contact the respondent directly or indirectly through telephone, than it is through the questionnaires he can be contacted. Questionnaire is a set of questions to which a respondent response. This questionnaire are made available to respondent either by enumerators or by mailing them. The respondent is requested to return them dully filled in within a reasonable time period.

This method is normally used in survey and before adopting this a pilot testing is conducted. The success of this method is based on proper and careful questionnaire, so as to get required and relevant information.

D. By Schedules

There is hardly any difference in schedules and questionnaires in relation to their structure. Both are the set of questions for the purpose of response to be taken from respondent. The difference lies only in method of sets getting filled in.

In questionnaire respondent himself record his answers to the question as per his understanding. Where as in schedules the enumerators are made responsible to record the answers given by the respondent.

7. Project Execution :

The project execution is very important stage in the research process. It should be executed in a systematic manner. Therefore periodical and occasional check is essential for verification of the data collected.

A careful watch should be kept on the authenticity of the collected data. It should also be verified to find out whether respondent has given the proper response. The efforts should be made to organize the proper response by the respondent.

8. Data Analysis

In research process after the data is collected it is required to analyze the meaning of objective. The collected data is processes through various operations. So in order to make raw data meaningful these operations will help us to draw proper conclusions.

The raw data will be processes by establishing categories, through coding and tabulation. Before the advancement of computers all these processing operations were conducted manually. Now with the technological advancement all these operations are conducted through computer applications.

Now researcher who want to analyze the collected data with the help of certain statistical measures, can make the use of computers.

9. Testing of Hypothesis

It is only after the analyzing data, researcher can go further in testing of his hypothesis, in which he can ascertain the facts to support or disagree with the hypothesis.

There are various methods of testing of hypothesis, researcher can select any of them depending upon the subject matter of his research study. It may be possible that in some cases the researcher does not have hypothesis in the beginning. So the general assumptions on the basis of data can be considered as hypothesis, which can be tested by the prospective researchers.

10. Data Interpretation

After hypothesis is tested than researcher can go further for drawing of general inferences so that he can arrive at making of a statement. The ability of any research is largely based on its capacity of making general statement.

Researchers having no hypothesis in the beginning, can take help of the any established theory to explain the concept through a process of interpretation.

11. Report Writing

Report writing is final job to be done by the researcher. Research work which is conducted by him is finally presented in form of research report. Report must be written in systematic manner, which normally should have following parts.

A. Preliminary Part - which should include lists of tables, charts and graphs along with index of the work.

B. Main Text of the report – this will contain the statement of research problem, scope of the problem, research methodology used, limitation and period of the study.

It will also include logical sequence of the various chapters and its break down in paragraphs.

C. Conclusion Part - will consists of the observations made in the research study On the basis of these observations what ever conclusions are drawn are included here. Finally we sum up the recommendations or the suggestions of the study.

D. Reference Material – What ever references are used in the research should also be listed out properly. This reference material used in the research study may be in form of published or unpublished in nature. It may be in form of printed books, journals. Some times unpublished data is also used.

All theses material is used in support of the certain claims made in the research study. So the Appendices and bibliography are given here to justify the claims made.

Steps in Research Process :

(1) Deciding - What an I expected to do ?
- What do I already know.
- What question do need to answer.
- What I need to find out?

(2) Finding - What sort of Information I need where do I find this information.
- Should I use cataloges, index computers.

(3) Using - What do the resources tell me.
- Do they answer my question.
- Do I have skills to extract & use information.

(4) Recording - What do I need to make a role of ?
- Can I make effective notes.
- Can I organise my notes.

(5) Presenting - How I am going to pull this information together and present it so that I communicate effectively.

(6) Evaluating - I found information to complete the task.
- How well did I do.
- Am I satisfied with the result ?
- Do I need more practice with any stage or skill.

Basic in Research process :

Following Steps are the effective process for conducting research.

1. Choose your topic.
2. Find basic information.
3. Refine your topic.
4. Locate material.
5. Evaluate references of material.
6. Organising and report writing.
7. Composition and bibliography.

Figure of Flow Chart on Research Process

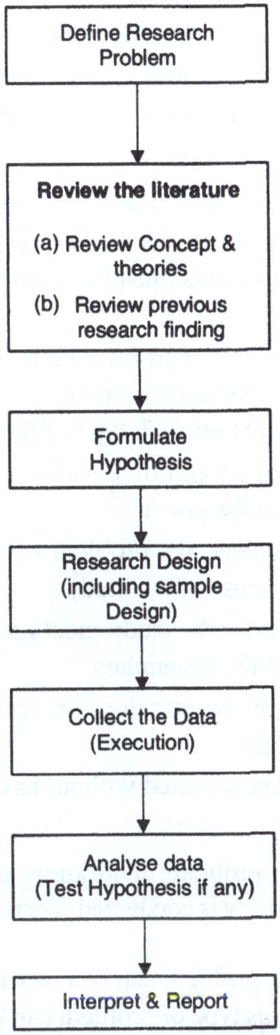

RESEARCH PROBLEM

The entire research activity is based on the proper identification of the research problem. Unless the researcher diagnose the problem correctly no appropriate solution can be provided.

Therefore research activities forms a cycle, which start with identification of research problem and ends on the solution to the problem. It means research problem is a axis, where whole research activities revolves. This research problem will speak out the aims and objectives of the research work.

Research problem is the situation that causes researcher to feel confused. It is the demarcation of the problem area within the context of the problem situation. It is the problem situation which may give rise to research problem.

The researcher who is conducting research study experiences some difficulties in relations to solutions to be sought. Therefore sources usually contribute to identification, selection and formulation of the research problem.

The research problem forms a cycle, as research start with problem and ends on the problem. The problem arises on account of difficulties in research faced by the researchers and he has definite objectives to be attended by the research.

As there are various alternative available for research and researcher has to opt for one. In this case he is totally undecided about selection of the alternatives.

Criteria for Selecting Research problem

Researcher should take utmost care in selecting research problem. It should be his own independent thinking in relation to the problem. The necessary guidance can be sought to work out the problem and following facts must be given due consideration while selecting research problem.

1. The subjects which are overdone and on which sufficient work has already been conducted in particular area should be normally avoided by the researcher. It will be very difficult to throw new light on such type of the topics.
2. There are certain topics which are controversial in nature and in spite of our all efforts no purposeful conclusions can be drawn.
3. While selecting research problem, the problem which is very narrow or very vague should be avoided for the purpose of the study.
4. What ever subject is selected for the study must have the resources available, which should be within the reach of the researcher.
5. Selection of research problems require due consideration for time, training and cost involved in the research project.
6. No research problem should be selected without having primary study of the subject.

Identification of the Problem

The desire to solve research problem or to know more about the facts is the main motivation for which the research study is conducted. Therefore it is necessary for researchers to identify the research problem.

The identification of research problem can be classified as under

A. Conceptual Problem - This type of problem can be solved by creative thinking for the problem.

B. Empirical Problem – This type of problems can be solved by inductive reasoning, which should be based on the observation made.

C. Logical Problem - This problems can be solved by deductive methods.

Now it is very clear from the above expiation of the facts that identification of the research problem means clear and accurate assessment of the research problem. In order to solve the research problem.

Identification of Research problem in Business

There are certain business problems which require critical assessment of the certain business documents. Evaluation of these documents will provide us the base for taking appropriate decision.

The following methods are being used in the business to locate the business problem and assessment of the facts provide the solution to the business problem.

1. Review of the business reports and office records of the organization.
2. Observation of the business conditions.
3. Discussion with the competent office executives.
4. Careful observation of the office records of the successful units.
5. Critical assessment of the published records.

Selection of the Research Problem

The problem which is selected on specific rational helps us to complete the project in stipulated time period and within the prescribed financial outlay. In order to boost up the moral of the researcher proper selection of the problem is essential.

The factors which requires due consideration for selecting research problems are stated as below :

1. Interest of the researcher

The interest of the researcher is the main factor while selecting problem. It is the attitude, spirit and dedication of the researcher towards his research study, that will generate the interest in the research study. Therefore the personal interest of the researcher is very much essential in any of the research study.

2. Importance of the topic of the study

Topic selected for the study should not only be socially relevant but it must be useful for the society. It should be having the reference to the practical field. The results of this type of social relevant study will be used by the concerned individual or group of the individuals.

3. Avoid the selection of the monotonous problem

The researcher should avoid to use monotonous problem for the purpose of the research study. Such topic for the research may not be able to generate solution for utility of the subject matter. Therefore there should be novelty of ideas in the subject so as to have the proper utility of the research study.

4. Resources for the Research

The availability of the resources is very important factor in the research study. These resources are related to financial and technical assistance to the researcher along with his own experience in this respect. Resources has it's application in the research within the prescribe time period. The efforts in this direction should be made for having proper application of the resources within specific time period.

5. Data Availability

The data required for the study should be easily accessible. In case if researcher decides to make use of the primary data then it is necessary to ascertain the response of the respondent. If secondary data is to be used then authenticity of the data is to be tested before it is put to an application. Proper care should be taken to collect sensitive data which is required for the purpose of the study.

Steps in Selecting Research Problem

1. Statement of Problem

Researcher is required to explain his problem under the study. The explanation to his problem is known as statement of the problem. First problem should be explained in the general manner. This will provide the clarity of understanding of the problem and then only researcher will be able to state his problem properly.

Even before deciding the problem for the study, pilot study can be conducted. When researcher makes general description of his subject there may be lot many ambiguities in the beginning, which requires rethinking of the subject.

So the problem stated should be suitable for necessary changes or if needed it can be redesigned.

2. Nature of Research Problem

After defining the research problem and making proper statement of the problem, it is necessary to understand the origin of the problem for it's clarity. In order to make topic of the research further more clear researcher can undergo in discussion on this topic with the persons having good knowledge of the similar problem. With the help of this discussion problem can be properly assessed.

3. Review of the Available Literature

What ever literature is available in relation to the proposed study,it must be carefully examined before selecting problem. This exercise will be very useful which can provide many suggestions and later on required additions can be made in the research problem.

4. Discussion to enrich the subject matter

The research topic should always be open to the discussion with experts. It is through this discussions only new ideas can be developed. This is known as experience surey. The discussion should be healthy one which must lead to general approach to the problem. This should also lead to enrichment in the subject matter of the study.

5. Re-designing of the Research problem

Research study requires the ability of the researcher to put his research ideas on the right track then only he will be able to convert his ideas in to working hypothesis. This will make his research problem operationally viable. This will create the base for finding out working hypothesis for the research study.

6. Utility of Research

Research studies are conducted not only for academic satisfaction of the researcher, but it is also conducted for utility of the society. The findings of the research study should reach to the concerned section of the society. Therefore research study with social relevance and utility should normally be selected for the study.

In order to provide the benefit of the research to the society, it's findings should be made public by way of publishing the results of the research.

7. Practicability of the Research Study

The factor of the practicability should be considered well before the research problem is selected for the study. Unless the feasibility of the study is properly tested research problem can not be selected. Personal whims and unrequited imagination of the researcher will spoil the practicability of the study. There should be rationality and manageability in the research study.

Normally research study is tested and judged by the contributions of the researcher, observations made by him in relation to the study conducted and his analytical approach for the study.

QUESTIONS

1. What do you mean by Research Methodology? Explain it's significance and describe compile the different types of Research.
2. Explain the steps in Research Process with the help of Flow-Chart of the Research Process.
3. Describe the different types of research, clearly pointing out the difference between an experiment and a survey.
4. "Creative management, whether in public administration or private industry, depends on methods of inquiry that maintain objectivity, clarity, accuracy and consistency." Discuss this statement and examine the significance of research."
5. "Research is much concerned with proper fact finding, analysis and evaluation." Do you agree with this statement? Give reasons in support of your evaluation."
6. Write shot notes on (any two)
   (a) Criterion of good research. (b) Dependent and Independent variable.
   (c) Case study method. (d) Components of a Research Problem.
7. Write the steps used in selecting a research problem.

■■■

CHAPTER 2

RESEARCH DESIGN

After going through this chapter, you will be able to know :

- Introduction of Research Design
- Definition of Research Design
- Classification of Research Design
- Features of Research Design
- Factors for Appropriate Research Design
- Types of Research Design
- Design for Exploratory
- Selection of Research Problem
- Steps of Research Design
- Features of Good Research Design
- Criteria of a good Research Design
- Uses of Research Design

After the identification of research problems the preparation of research design followed subsequently. The research design is concerned about the mechanism as to how the research is to be conducted. It is a procedure regarding the collection and analysis of data, relevant to research work by appropriate and economic method.

Research design means a process by which researcher will be in a position to understand the structure of the research and the various steps to be taken in the process of research. Researcher is supposed to understand the type of data required for the research. The area of the study and the time frame during which research is required to be completed. As such it naturally indicate outline of the research work and its operational application.

Definition

Research design has been defined by different social scientists in different terms.

Vimal Shah has said that "The design is the plan of study and as such it is planned in every study uncontrolled as well as controlled and subjective as well as objective".

To quote P. V. Young "The design results from controlling general scientific model into veried research procedure".

According to E.A. Suchaman "Research design represents a compromise dicted by the many practical considerations that go into social research.... A research design is not a highly specific plan to be followed without deviation, but rather a series of guide posts to keep one in the right direction".

As Selluz, Jahoda, Deutsch and Cook state "A research design is the arrangement of conditions from collection and analysis of data in a manner that aims to combine relevance to the research purpose with economy in procedure".

In the words of *R.L. Achoff* "The idealised research design is concerned with specifying the optimum research procedure that could be followed where there is no practical constraint".

Miller has defined "Designed research" as the planned sequence of the entire process involved in conducting a research study".

A few other definitions of research design are "Design helps the investigator obtain answers to the questions of research and also helps him to control the experimental extraneous and error variances of the particular research problem under study".

"A research design designates the logical manner in which individuals or other units are compared and analysed, it is the basis of making interpretations from the data".

"Research design is the plan, structure and strategy of investigation conceived so as to obtain answers to research questions and control variance".

"The challenge of a research design is to translate the general scientific model into a practical research operation. Research design will refer to the entire process of planning and carrying out a research study".

The other important factors will naturally be related to sample which will be needed in research work and method to select such samples is related to the collection and analysis of the data. Taking in to consideration the above referred desicions the research design can be classified as under :

Classification of Research Design :

1. **Sampling design :** It relates to the method of selecting items for the purpose of study.
2. **Observational design :** This relates to the conditions under which observation are made.
3. **Statstical design :** This relates to the information and data which is to be collected and is further required to be analysed.
4. **Operational design :** This deals with certain techniques carried out in the process of research.

Features of Research Design

1. Research design is a plan which specifies which type of infomation will be relevant to the research study.
2. The collection and analysis of the data requires the strategy to be specified which will be used for research study.
3. Research design has main emphasis on time period involved and financial outlay required. The research activity may suffer if these aspects of time and finance are developed as constraints.

   Therefore research design has to have:-

   (a) Proper understanding of research problem.

   (b) Data collection procedure

   (c) Population to be studied

   (d) Data analysis methods.

   Why Research Design

The research design has its own importance in the process of research methodology. It is required for the following reasons.

(i) For smooth conduct of the research activity.

(ii) For relevant collection and analysis of data.

(iii) It provides strong and firm base for the research work.

(iv) It helps researcher to organise his ideas in relation to research activities and it also helps to avoid flow and inadequacy.

(v) In absence of proper research design comprehensive review of the research study is impossible.

P. V. Young presents a summary of the main contents of an ideal Research Design :

(a) Sources of information to be tapped.

(b) Nature of study.

(c) Objectives of study.

(d) Sociological context of study.

(e) Geographical areas to be covered by the study.

(f) Time period to be encompassed.

(g) Dimensions of the study.

(h) The basis for selection of data.

(i) Techniques to be used for gathering data.

(j) Specification of time schedule for completing the research.

All these features are to be present in a Research Design. While formulating a research design a researcher has to take decisions with respect to the following :

What is the study about and the types of data that are needed?

Why is the study being made?

Where can the needed data be found?

Where or in what areas, will the study be carried on?

When, and for what period of time will the study be conducted?

How much materials or how many cases will be needed?

What techniques of gathering data will be adopted?

Decisions on all these questions lead to the formulation of a research design, which are put into black and white by the resarcher before he really enters into the universe of his research project for data collection and analysis.

Factors for appropriate research design

The following factors are involved in appropriate research design.

(1) The methods for obtaining necessary information for the research study.

(2) The ability and skill of the researcher and his associates in conducting the research study.

(3) The objectives of the research problems are required to be properly studied.

(4) The nature of the research problems also requires due consideration.

(5) The avilability of funds and the time period available are the important factors for the purpose of research study.

It is necessary to understand various concept of research design for conducting research study.

A. Variables

Dependent and Independent Variables means different quantitative values such as height, weight etc. The qualitative phenomenon is also important on the basis of their presence or absence.

This phenomenon can be continuous or non continuous.

(1) A continuous variable assumes a numerical value within a specific range. For example if we collect data in respect of children it is noncontinuous variable.

(2) The variables explained above can be dependent or independent.

If one variable depends for its causes on other variable, the other variable is independent variable. For example income and demands. Demand of the product is based on the income of the consumer. In this case income is an independent variable and demand is dependent variable.

B. Extraneous Variable

An independent variable sometimes may not be a part of the study eventhen they may affect the dependent variable which is the part of the study. So such independent variable not concerned with the study but affecting dependent variables are known as extranuous.

For example- demand of luxrious items like car - In this case income of consumer is one variable and demane of car is another variable. In this case 'Income' is independent variable and 'demand of car' is dependent variable. Here the requirement of car can be affected by the loan facilities provided by the bank. Loan facility is independent variable which is not a part of the study but it affects the dependent variable of demand of car purchase. Hence the loan facility is called "extranuous variable."

The research study should be so designed that the effect upon the dependent variable should be on the basis of independent variable related to the course of study where as such dependent variable should not be affected by extraneous variable.

C. Restrain on extraneous variables

It is clear from the above fact that extraneous variable may affect the dependent variable of the study. Therefore it is very much required to control and minimise the influence of extranuous variables from the study.

D. Confounded Relationship

When it is not possible to avoid or to minimise the influence of extraneous variable on dependent variable and it is not free from such influence then the relationship between the independent and dependent variables is called as confounded relationships.

E. Hypothesis-Relationship

It is a preassumption on which research study is based. It is a prediction made by the researcher when he starts the study. Hypothesis must contain both variables ie dependent and independent one. Such study must be related to relationship between these two variables. When this hypothesis relationship is tested by scientific method it is termed as hypothesis research.

F. Testing of Hypothesis

When th object of research is to test hypothesis relationship it is called as hypothesis testing. If the independent variable is manipulated in the research it is called as experimental hypothesis testing. Where the indenpendent variable is not manipulated in the study it is called non-experemental hypothesis testing.

For example- If researcher wants to study whether intelligence affects the selection in process of recruitment. 10 students are selected for study. Test the relationship of intelligence and selection by calculating coeffectient of correlation between the two. In this non-exerimental hypothesis testing variable of intelligence is not manipulated.

Now 10 student are divided in to two group of 5 in each group A and B. Group A is provided special coaching and training techniques by expert. Group B is exposed directly without any training. At the end of the programme the performance of the groups is judged. This is a case of experimental hypothesis tesing where the independent variable of intelligence was manipulated.

G. Testing Group in conditions

If group is provided with normal conditions in testing it is called as control group and when group is provided with special conditions it is called as experimental group.

H. Testing conditions

Both the groups ie experimental and control are put to different testing conditions and are termed as treatments. If we want to test impact of application of various fertilizers on crop of cotton. So the application of two varieties will be treated as two treatments.

I. Testing process

The process of testing or examining the facts of hypothesis in relation to research problem in called as experiment. This experiment can be classified.

(a) absolute or (b) comparative

If we study impact of one fertilizer of cotton productivily. it is abslute experiment and of we study impact of various fertilizers on cotton productivity it is comparative experiment.

J. Testing Units

The one decided sets on which different testing condition or treatments used are called as testing units or experemental units. It is necessary to take proper care for selecting such testing units.

Types of Research Designs

The type of research study differ from the nature of the study to be conducted. Therefore different types of research design are used in case of exploratory descriptive or testing of hypothesis research studies concerned.

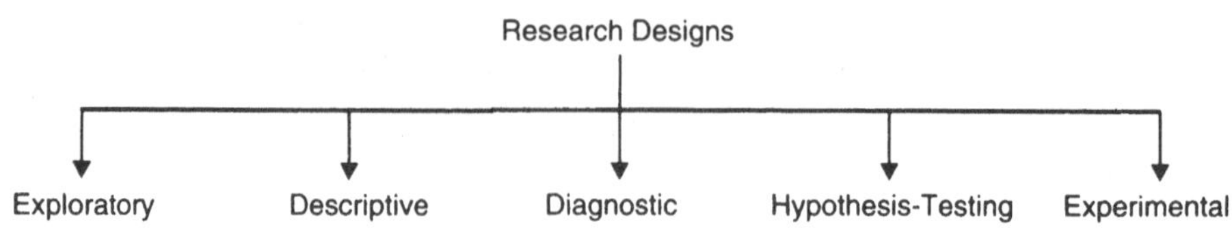

(A) Exploratory research design-

The basic purpose of the exploraling research studies is to achieve new insight for formulating research problem or to develop hypothesis for the purpose of further studies.

These studies are more important in case where very little information is available. The following methods are used in context of this research design.

1. The survey of the literature related to the field.
2. The survey of the people who have an experience of the concerned problem.
3. The insight stimulating examples related to the study are analysed.

Design for Exploratory or Formulative Studies

Exploratory studies have, in the main, the purpose of formulating a problem for more precise and structured investigation or of developing hypotheses. An exploratory study may, however, have other functions too, e.g., increasing the investigator's familiarty with the phenomena he wishes to study in a subsequent, more structured investigation or with the setting in which he plans to carry out such an investigation. An exploratory study may also serve as a basis for clarifying concepts, establishing priorities for further research, gathering information about practical possibilities for carrying out research in specific real-life settings, etc. "Exploratory studies," says Katz, "represent the earlier stage of science." From its findings may emanate the knowledge that helps the researcher in formulating a problem for research or in developing hypothesis to be tested subsequently.

Let us try to understand to some satisfactory extent the nature of an exploratory study by an analogy. A doctor who is called upon to attend to a patient who is totally unfamiliar with him will ask him various questions concerning his complaints, will examine the various parts of the patient's body using different instruments at his disposal and persue the patient's pathological reports or records (if any) and so on. On the basis of this exploration, the doctor may find himself in a position to pose a question like, "could it be typhoid?" One of his hypothesis relating to the above question may be, 'It is typhoid'. The doctor's subsequent treatment in the nature of antibiotics will constitute a test of the hypothesis. If the patient responds favourably to the treatment there is room for believing that the typhoid hypothesis is tenable. If the post-treatment observations suggest unfavourable response, the typhoid hypothesis is falsified. Such a test of hypothesis does not belong to the realm of exploratory studies.

The above example illustrates the nature of an exploratory study and also how it differs from the problem-solving and hypothesis testing studies. In the initial stages when the doctor was asking the patient all manner of questions and was examining him, using various instruments and scrutinizing various reports, the doctor was simply exploring i.e. conducting some sort of an exploratory study. The end-result of this exploration was the question (problem) that suggested itself the set of alternative hypothesis.

(B) Descriptive research design

Descriptive research aims at protecting accurate character of particular individual of a group as it describes their characteristics.

This study can be conducted in factory for the workers like cart group. and educational level. or age and income group etc.

This study is primarly conducted with an object to know their attitude towards anything. The research design for this study involves forming of an objective, selecting sample, collection and analysing of data and the result.

(C) Diagnostic research design

The diagnostic research studies decide the frequency of occurence in association with something else. Cause-effect relationships are described in such studies. Its a fact finding exerscise. This includes screening, classification, description and prediction act of problem reported.

It is a discovery of relevant variables that are associated with it in varying degrees. Its a typical problem solving strategy. The relevant data for the purpose of the study Can be ontained from observation and formal testing. This includes the steps like applications of statistical computations.

Main Difference in this two research design

Research for Design points	*Expdoratory*	*Describtive/Digonostic*
(I) Overall	Flexible	Rigid
(II) Sampling	Non-Probability	Probability
(III) Observational	Unstructured	Structured
(IV) Statistical	Unplanned	Preplanned
(V) Operational	No fixed decision	Advance decision.

(D) Experimental Pre research (Hypothesis testing research design)

Experimental research is related to laboratary processing. By this method researcher is in a position to observe the effects of manipulation of independent variable to dependent variable.

This study is more useful for testing of causal hypothesis. The basic purpose of this study is to test causal relationship between variables. The validity of experimental studies depends on equivalence between control group and experimental group. These group should not differ in significant respect except by chance.

Then only they can be compared in terms of assumed effect by comparing and reandomization in which we can have better estimation of experimental error.

The main purpose of the research design is to set up research project which must end to logical conclusion for which following stages are adopted.

(E) Experimental research design

Basic Principles of experimental designs

The basic principles of experimental designs are randomization, replication and local control. These principles make a valid test of significance possible. Each of them is described briefly in the following subsections.

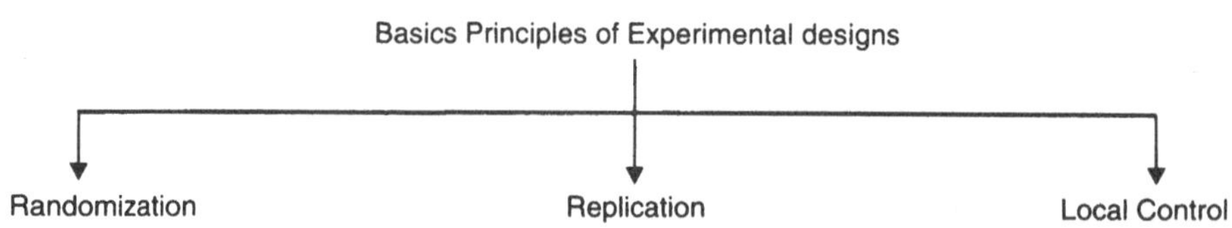

(1) **Randomization.** The first principle of an experimental design is randomization, which is a random process of assigning treatments to the experimental units. The random process implies that every possible allotment of treatments has the same probability. An experimental unit is the smallest division of the experimental material and a treatment means an experimental condition whose effect is to be measured and compared. The purpose of randomization is to remove bias and other sources of extraneous variation, which are not controllable. Another advantage of randomization (accompanied by replication) is that it forms the basis of any valid statistical test. Hence the treatments must be assigned at random to the experimental units. Randomization is usually done by drawing numbered cards from a well-shuffled pack of cards, or by drawing numbered balls from a well-shaken container or by using tables of random numbers.

(2) **Replication.** The second principle of an experimental design is replication; which is a repetition of the basic experiment. In other words, it is a complete run for all the treatments to be tested in the experiment. In all experiments, some variation is introduced because of the fact that the experimental units such as individuals or plots of land in agricultural experiments cannot be physically identical. This type of variation can be removed by using a number of experimental units. We therefore perform the experiment more than once, i.e., we repeat the basic experiment. An individual repetition is called a replicate. The number, the shape and the size of replicates depend upon the nature of the experimental material. A replication is used:

(i) to secure more accurate estimate of the experimental error, a term which represents the difference that would be observed it the same treatments were applied several times to the same experimental units;

(ii) to decrease the experimental error and thereby to increase precision, which is a measure of the variability of the experimental error; and

(iii) to obtain more precise estimate of the mean effect of a treatment, since $\frac{\sigma^2}{y} = \frac{\sigma^2}{n}$, where n denotes the number of replications.

(3) **Local Control.** It has been observed that all extraneous sources of variation are not removed by randomization and replication. This necessitates a refinement in the experimental technique. In other words, we need to choose a design in such a manner that all extraneous sources of variation are brought under control. For this purpose, we make use of local control, a term referring to the amount of balancing, blocking and grouping of the experimental units. Balancing means that the treatments should he assigned to the experimental units in such a way that the result should be collected together to form a relatively homogeneous group. A block is also a replicate. The main purpose of the principle of local control is to increase the efficiency of an experimental design by decreasing the experimental error. The point to remember here is that the term local control should not be confused with the word control. The word control in experimental design is used for a treatment. Which does not receive any treatment but we need to find out the effectiveness of other treatments thorugh comparison.

Important Experimental Designs

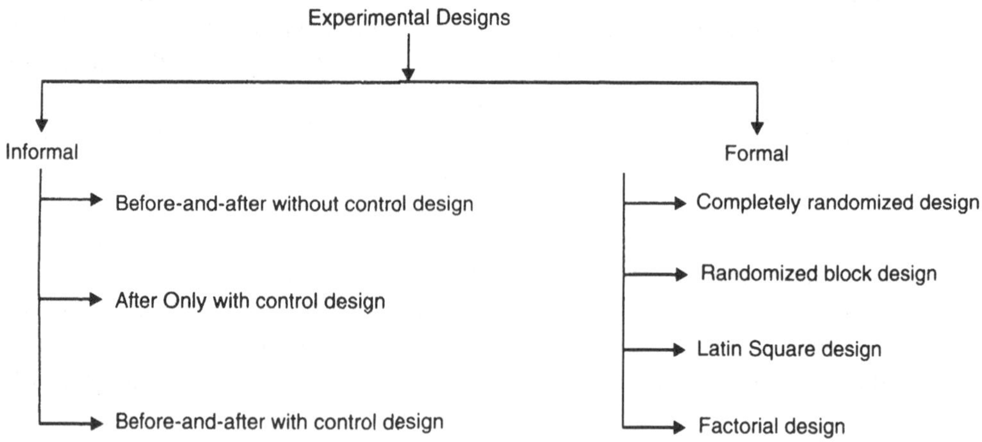

There are several designs:
I. Informal Experimental Designs
II. Formal Experimental Designs

I. Informal Experimental Designs

(a) Before-and-after without control design
(b) After-only with control design
(c) Before-and-after with control design

Before-and-after without control design

A single test group or area is selected and the dependent variable is measured before the introduction of the treatment in the "Before and after without control design". The dependent variable is again measured after the treatment.

After-only with control design

In the "After only with control design" two groups or areas (test area and control area) are selected and the treatment is introduced into the test area only. The dependent variable is then measured in both the areas at the same time.

Before-and-after with control design

In the "Before-and after with control design", two areas are selected in which the dependent variable is measured for an identical time period, before the treatment. Then the treatment is applied into the test area so that the dependent variable can be measured in both for an identical period of time after the treatment.

II. Formal Experimental Designs

(a) Completely Randomized design
(b) Randomized block design
(c) Latin Square design
(d) Factorial design

(a) Completely Randomized Design

Completely Randomized Design (CRD) is a simple research design. Subjects are assigned at random for experimental treatment. Principle of replication and principle of randomization of experimental designs are involved in this type of design and random replications design are the two forms of CR design. In the former a sample is selected at random from a defined population. In the latter a number of repetitive treatments are given in each case minimizing the effect of differences.

(b) Randomized block design

Randomized block design (RB design) makes use of the principle of local control. Subjects are first divided into groups (blocks). Subjects are relatively homogeneous in regard to selected variables in the RB design.

(c) Latin square design

When two extraneous factors are applicable, Latin square design is effective. Out of the many items; say for example, five items are placed in five places each in the square but used only once in each row and column. The two blocking factors on two sides of the square (vertical and horizontal) are placed. It is generally used in agriculture.

(d) Factorial designs

When a large number of factors affect a particular problem, factorial designs are used. There can be simple factorial designs and complex factorial designs. Generally two factors are involved in the former, and many factors (multi factors) in the latter. The independent variable to be manipulated is called experimental variable, and the extraneous variable to be controlled by homogeneity is called control variable.

Selection of research problem

The research problem may be selected from the following sources :

1. Theory
2. Every day problems
3. Technological changes
4. Unexplored areas
5. Discussions with supervisor

The selection of research problem depends on whether research is being conducted for obtaining a degree or it is for academic interest. If the research is for Masters degree level, M.Phil. or Ph.D., the problem may be more specific and limited in scope and may offer itself for completion within a specified time. On the other hand if the research is for academic interest, time should not be a binding factor. Here the sole objective is to enrich the knowledge through the application of advanced research methods special assignments, research reports and articles may suggest some additional areas of needed research.

Steps in Research Design

A research design usually comprises of the following major steps.

A. Formulation of the objective of the study (why the study is conducted and means to indentify the problem and problem area).

B. Designing the method of data collection (what should be the technique of data collection).

C. Selecting Sample.

D. Collecting data - availability and its time period to findout
E. Review of the literature
F. Hypothesis selection
G. Designing of the experiment.
H. Data analysing process.
I. Report on findings of the study.

The above stages are normally, seen in any of the research design. In order to make design purposeful action the entire activity of design is required to be carried out with proper sequence right from indentifing problem to the end of design process report writing. In relation to the stages explained above, plan of the research design can be undertaken in the following steps.

1. Title of the Research

The title of the research should be properly worded. It should throw the light on the entire activity to be conducted. It should be easy for anybody to draw proper meaning of the proposed study to be undertaken.

The title so selected for the study should not be baised. It should not be emotional. It must be scientific in approach to its indication of the subject matter.

2. Statement of Problem

It is an expression of the title statement which relates to explanation of the problem. It should contain main aspect of problem identification and specify the research area. This should include scope and limitation of the study. This should be related to the object of the study by providing earlier studies conducted. Plan for the study should be properly defined in this stage.

3. Review of the literature

The researcher must review existing available literature with an object of developing hypothesis from it. In some research studies hypothesis may have been stated by previous researchers. In this stage stock of various hypothesis should be evaluated for its usefulness. This review of existing literature will provide base for proposed study insight for the study which will make researcher able to formulate correct plan for investigation.

4. Sources of Information

This indicates the outline of methods and procedures for collection of data. This may be from primary or secondary source. The design should indicate the choice of the type of data to be collected based on the nature of the study.

The out come of the study is totally dependent on proper and reliable data. The methods to collect such data is required to be specified in this stage.

5. Nature of Study

The researcher must specify the nature of the study, whether it relates to stastistical experimental or comparative or combination of any of these.

Defining proper nature of study will reduce the complexity of the problem and will lead to accurateness in the study.

6. Object if the study

The object of the study should be stated in clear terms. With specific goals to be attained the research study should aim at deseriptive study exploratory study or testing of hypothesis. This should be clearly, worked out along with relationship of this study with previous research study.

7. Socio-cultural context of the study

The utility of the study is proved for its socioculture context. If study is conducted with reference to investigation made which relates to the human behaviour of the person to be ascertained. The usefulness of the study will be based on now they as here, deviate or withdraw in relatise to their behaviour for the purpose of the subject under study.

8. Geographical Area

The research study is confined to specific geographical area in which study is fo be conducted. The area of the study may block, destrict, region devision state or nation depending upon the nature of the study.

9. Time Dimension of the study

The research study can not be conducted for any unlimited period. The success of the research study is based on reachable time period selected. It may be ranging for 5-10-15 or 20 yrs.

The study can also be confined to specific area i.e. pre independence or post-independent period British period mugal period etc.

This will help researcher to attribute the conclusions related to specific period.

10. Assumption of the Study

Basic assumptions of the studies should be clearly and properly stated. This may be in relations to defining new concept. This also specifies the terms to be used. It means common word usage having recurrence of its applicability. The interpretation is largely based on explaination to these terms in relation to the research study.

11. Basic of Selecting Data

This relates to the choice of samples to be selected for the purpose of the study. The proper care should be taken regarding the choice of the samples.

1. It must represent the universe.
2. It must be unbaised.
3. It must be adequate in size to produce results.
4. It must be designed to be effective.

12. Techniques of the study

The preparation of the outline of the research work decides the suitable techniques for collection of data. Which of the method will be used in this reference is required to be specified.

It is possible that only observation method may be used for information, or it can be collected through questionnaire or even both the methods can be commonly used or any other method is likely to be used should be specified here.

13. Control of Error

This is applied in experimental study. The researcher must know what variables are operating in the given situation. It is possible to control error in the laboratory or in field investigation. Normally key variables can be controlled. In all the cases it in necessary that variables and their control needs to be described.

14. Establish the reliability and validily of Instruments

The reliability and validity of instrument of data collection is necessary in case of impartial study unless these factors are established it is very difficult to evaluate such data collected.

15. Scheme of Chapterisation

It is part of the entire research study that there should be proper plan of study to be classified into various chapters.

According to the sub classification and head of particular chapter entire study will be divided in various chapter right from introduction to the final chapter of conclusions.

16. Development of Bibliography

The bibliography refers to refsource of the study material used. If reference can be made to use of published and unpulished books, research papers articles reports or study conducted All these reference should be properly mentioned in the study in relation to their author, pubblisher, year of publication and page reference.

Features of Good Research Design

The good research design should have the following considerations.

1. It must be minimum bias and must have maximum reliability.
2. The research design must have flexibility. It should contain discovery of ideas and insight as required in case of exploratory studies.
3. In case of disceriptive studies there should be accurate description of the situation.
4. The reliability of the evidence collected is considered a good research design.
5. It should involve testing of hypothesis.
6. The good design should permit inferences about the causality.

Criteria of a good research design

A good research design should be of definite help in achieving optimum objectivity, reliability, validity and generalisation.

1. Objectivity

When a phenomena is observed in its true form without being affected by observer's own views it may be termed as objective observation. For example when we say that milk is white it is an objective statement. But if we say that - milk is the most useful drink - the statement may not be purely objective. The objectivity of the findings pertains to either collection of data or scoring of the responses or both. It may be judged by the degree of agreement between the final scores assigned to different individuals by more than one independent observer. Any research design should permit the use of measuring instruments which are fairly objective in which every observer or judge seeing a performance arrives at precisely the same report. Objectivity seems very easy to achieve but in actual practice it is not so.

2. Reliability

Reliability refers to consistency throughout a series of measurements. The investigator should frame his items in such a way that the respondent cannot but give only one genuine response. That is to say, if a respondent gives out a response to a particular item, he is expected to give the same response to that item whenever he is asked subsequently. Check items, administering the same test repeatedly, using a series of parallel forms are methods used in determining the reliability of the responses given out by a respondent.

3. Validity

The researchers must make sure that any measuring instrument selected by him is said to be valid when it measures what is purports to measure. For instance a originality test, constructed for measuring originality should measure only originality and nothing else. There are a good number of procedures for establishing the validity of test. Some such procedures are : validating the present data against a concurrent criterion or a future criterion on a theory etc.

4. Generalisation

Most research is concerned not only with the effect of one variable upon another under the particular setting studied, but also with its effect in a natural setting and on a larger population. That is how best the data collected from a sample can be utilised for drawing certain generalisations, applicable to a larger group (population) from which the sample is drawn. However, complete generality is only a myth and it is true only under given conditions.

Uses of Research Design

1. The main use of the research design is that it provides the answers to various questions while conducting the study:
   a. What is object of the study?
   b. What data is needed for the study?
   c. How these data so required for the study will be collected?
   d. What will be the time, place and sample of the study?
   e. How will the data be analyzed and interpreted?
   f. What will be the cost, time and labour needed in the study?
   g. What will be the level of accuracy desired?
2. It acts as a standard and guidepost.
3. It helps in carrying out research objectively, accurately and economically.
4. It acts as an insurance against future failure.

QUESTIONS

1. Explain the concept of Research Design. Describe some important research design for research study.
2. How do you define a research problem? Give three examples to illustrate your answer.
3. Explain the meaning and significance of a Research design.
4. Describe some of the important research designs used in experimental hypothesis-testing research study.
5. "Research design in exploratory studies must be flexible but un descriptive studies, it must minimise bias and maximise reliability." Discuss.

■■■

CHAPTER 3

SAMPLE DESIGN

After going through this chapter, you will be able to know :

- Introduction of Sample Design
- Factors affecting the size of the sample
- The Role of Sampling Theory
- Stages in Sample Design
- Sampling Procedure Criteria
- Sample Design Characteristics
- Types of Sample Design
- Probability Sampling
- Double Sampling
- Comparision of Probability Sampling Design
- Use of Normal Distribution
- Selection of Random Sample
- Problems

A Simple design is a definite plan for obtaining a sample from a given population. Population or Universe means all the items of the inquiry or investigation made are completely enumerated in inquiry. Such a inquiry is conducted as census by the Government. This type of inquiry requires large number of manpower and great deal of time and money.

This type of work is very rare as we know the Government itself conducts such type of censes work once in ten years. There are lot many difficulties in conducting study of universe as it is impossible to examine each and every item of the universe.

Where study is conducted for small universe there is no necessity for any sample. When studies are conducted in relation to choice of consumers, behaviour of society then cost and time period factor forces us to select few items from the universe. Such representative factor of the universe is called sample.

Sample design may as well laydown that number of items to be included in sample i.e. size of the sample. The selected respondents constitute what is known as sample.

The process of the selection of sample is called sampling technique. The survey so conducted is known as sample survey.

Sample design should be decided well before the date for the research study is collected. It is excepted from the researcher that he must select design which should be reliable and appropriate for the research study.

Factors Affecting the size of the sample

The size of the sample depends upon a number of factors, the chief of which are stated below.

1. Homogeneity or heterogeneity of universe

If the universe is comparatively homogeneous a smaller size of the sample may be sufficient. If all the units were exactly alike one single unit could serve as sample, but if the universe is heterogeneous very few units are similar and thus the sample has to be essentially large in size.

2. Number of classes proposed

If a large number of classes are to be formed the sample must be large enough so that every class may be of a proper size suitable for statistical treatment. If the size of the sample is small there may be some classes which may contain one or two units only. Some may remain totally unrepresented. The result is that they can not be analysed properly and the generalisation based upon them will also not be correct. Thus large the number of classes proposed greater will be the size of the sample.

3. Nature of study

The size of the sample will also depend upon the nature of study. If an intensive study is to be made continuing for a pretty long time, large sample is unfit for the purpose, as it will require very large finance and other resources. Thus, in case of opinion surveys, where people have to be contacted for knowing their opinion about the problem under study the interview is quite short and much larger number of units may be included in the sample. Similarly in case of general study large number of cases can be taken but if the study of a technical nature, a large number may become difficult to manage.

4. Practical consideration

Practical considerations as availability of finance, time at our disposal, number of trained field workers etc. may also be taken as important factor in deciding the size of the sample. The limitations of these resources necessarily limit the size of the sample. One thing must be made clear at this tage. Although, these practical considerations do weigh heavily indetermining the size of the sample it should never be done at the cost of accuracy. Any amount of money, howsoever, small spent on an unrepresentative sample is pure and simple waste and must be avoided at all cost.

5. Standard of accuracy

It is generally considered that larger the size of the sample greater is the standard of accuracy or representativeness. Although, this is not true in all cases, as mere largeness of size is no guarantee for representativeness. A small but well selected sample may give better results than a larger and thoughtlessly selected sample. National Opinion Research Centre (N.O.R.C.) the premier opinion survey agency of America is now turning more and more to small size samples. In 1946 N.O.R.C. selected about 2500 cases in a standard poll. In 1948 it was reduced to 1300 cases only. As the technique of sampling is becoming more and more scientific and refined better standard of accuracy has been possible to be achieved by comparatively smaller samples.

6. Size of the questionnaire or schedule

The size of the questionnaire and the nature of questions to be asked is also a limiting factor for the size of the sample. Larger the size of schedule more complicated the questions to be asked, smaller is to be the size for proper administration.

7. Nature of cases to be contacted

The nature of cases to be contacted plays its own part in deciding the size of the sample. If the cases are geographically scattered a small sample is more suitable. On the other hand if the refusal rate is likely to be heavy or losses of cases likely to be quite big, a large sample has to be selected, so that after providing for the cases who may refuse to reply or those that may not be available due to other causes, the actual number of cases that is left out may be large enough to permit statistical analysis.

8. Type of sampling used

If absolute random sampling has been used a much larger sample is required. Random sample is reliable only when sufficiently large number of units have been selected, because it is only in large number of cases that law of statistical regularity properly works and every class of units gets a chance of being selected. On the other hand, if stratified sample has been selected reliability can be achieved in a much smaller size. But in stratified sampling the essential condition is that stratification must be proper. If stratification is unsuitable and improper a larger number will only add to the bias in the sample.

These are some of the broad considerations that are to be kept in mind. In fact no rigid number can be prescribed for an optimum size of the sample. The nature of the problem is the only deciding factor and the common sense and experience of the researcher the only guide in this matter.

The Role of Sampling Theory

This list of the steps in a sample survey has been given in order to emphasize that sampling is a practical business, which calls for several different types of skill. In some of the steps the definition of the population, the determination of the data to be collected and of the methods of measurement, and the organization of the field work - sampling theory plays at most a minor role. Although these topics are not discussed further in this book, their importance should be realized. Sampling demands attention to all phases of the activity: poor work in one phase may ruin a survey in which everything else is done well.

The purpose of sampling theory is to make sampling more efficient. It attempts to develop methods of sample selection and of estimation that provide, at the lowest possible cost, estimates that are precision at minimum cost recurs repeatedly in the presentation of theory.

In order to apply this principle, we must be able to predict, for any sampling procedure that is under consideration the precision and the cost to be expected. So far as precision is concerned, we cannot foretell exactly how large an error will be present in an estimate in any specific situation, for this would require a knowledge of the true value for the population. Instead, the precision of a sampling procedure is judged by examining the frequency distribution generated for the estimate if the procedure is applied again and again to the same population. This is, of course, the standard technique by which precision is judged in statistical theory.

A further simplification is introduced. With sample of the sizes that are common in practice, there is often good reason to suppose that the sample estimates are approximately normally distributed. With a normally distributed estimate, the whole shape of the frequency distribution is known if we know the mean and the standard deviation (or variance). A considerable part of sample survey theory is concerned with finding formulas for these means and variances.

One difference between sample survey theory and the classical theory of sampling is that the populations in survey work contain a finite number of units. The methods used to prove theorems are different, and the results are slightly more complicated, when sampling is from a finite instead of an infinite population. For practical purposes these differences in results for finite and infinite populations are seldomly important. Whenever the size of the sample is small (in terms of the number of primary sampling units) relative to the size of the population, results derived from an infinite population are fully adequate. In general, results for finite populations are presented in this book. In some of the more difficult problems, the theory for infinite populations is used to simplify the presentation.

Stages in Sample design

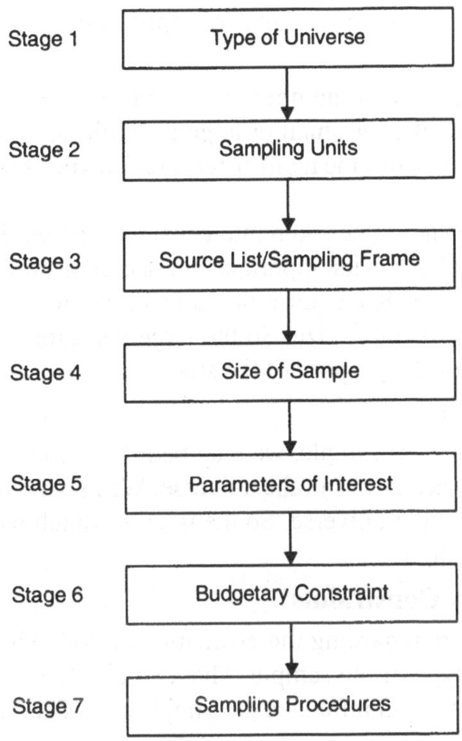

The emphasis should be given by the researcher on the following factors in order to develop sample design.

i. Type of Universe

The set of objections which are clearly defined is the basic stage in developing of any sample design. These set objectives are called universe. The universe may be certain or uncertain.

In case of certain universe the number of item can be decided and so we can have an idea about total number 6 items.

When such estimation can not be made it is totally uncertain universe.

Total number of industrial units in the region or number of bank's in the region is certain universe, but readership of news paper or viewers of T.V. news channel in uncertain universe.

ii. Sampling Unit

Researchers in supposed to decide sampling unit before sample to be selected. This sampling unit may be related to specific geographical area like National - State - Region

District - block - village or it can be economic group like H.I.G., M.I.G., L.I.G. BPL etc.

(Higher - Middle or lower income or below poverty line) Social grap - Upper cast lower cost.

Education group - primary - secondary - Higher Educational Institute.

It is the responsibility of researcher to select one or more such units for the purpose of study.

iii. Source list

In case of certain universe list in which information sampling frame is available is called source list which normally contain the names of all the items.

Researcher will be required to prepare the source test if not available. The list so available or perprepared must be representative of the universe.

iv. Sample Size

The size of the Sample means number of the items to be selected from the universe as sample. The size should neither be small or large it should be appropriate.

The sample size must fullfill the main criterion of flexibility, reliability, efficiency and representativeness.

The size of universe should be taken into consideration while deciding the size of the sample cost components also has its importance in deciding sample size.

The population variance also requires due consideration. If we take larger variance for study it will require bigger sample size. So the large variation in the universe will create large size of the sample for the purpose of the study.

v. Parameters of Interest

While deciding the size of sample, we may be interested in estimating the proportion of persons with some characteristics in the universe. We may be further interested to know the other factors relating to the universe. So the interest which we have in parameters will have impact on sample design.

vi. Financial Budgetary Constraints

The financial structure regarding the estimation of cost which have main impact not only on size but also on type of the sample. The availability of the funds required for the work and its proper estimate will influence the sample size. This further may lead to use the non - probability sample.

vii. Sampling Prodecure

It is the decision about the technique to be used in selecting the items from the sample. The researcher is required to explain his choice type of the technique he is going to apply because technique itself indicates sample design. Proper care should be taken by reseaarcher considering the aspects like size cost and error. According to the set size and cost the error should be minimum.

Sampling Procedure Criteria

The researcher must be aware of the fact regarding involvement of cost factor in sample design. One is in respect of the cost of the data collection and second regarding cost of wrong or incorrect inferences.

The incorrectness of inferences is major cause of worry therefore one must understand the reasons for such incorrect inferences. There are two main reasons for this.

A. Errors B. Biases

The differences occurring in a survey are classified as-

(a) Sampling errors
(b) Sampling Biases
(c) Non-Sampling Errors
(d) Non-Sampling Biases.

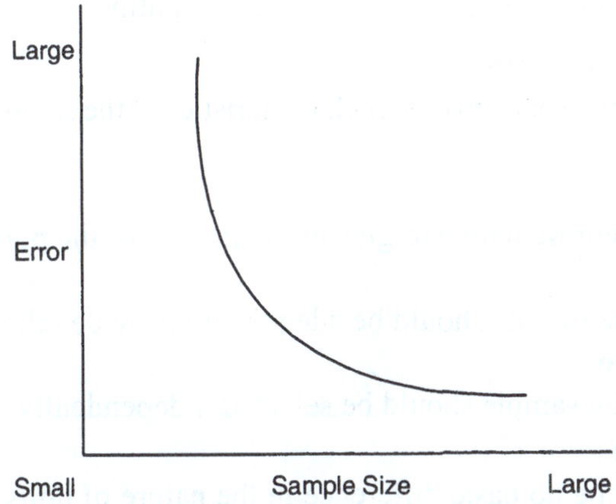

The relationship between Sample Size and the Sampling Error.

(a) Sampling Errors

Errors arising due to studying only a part of the total population are called Sampling Errors. These errors arise due to non-representativeness of the samples and the inadequacy of sample size. When several samples are drawn from a population, their results would not be identical. The degree of variations of sample results is measured by standard deviation and it is known as the standard error of the concerned statistic. As sample size increases the magnitude of the error decreases. Sample size and Sampling Error are therefore negatively correlated.

(b) Sampling Biases

The difference between expected value and 'True Value' of the parameter is the Bias. There are three reasons due to which the bias arises.

Following are the reasons:

(1) If the Sampling is done by a non random method.
(2) If the Sampling Frame is incomplete or inaccurate.
(3) If some sections of the population are not available or refuse to cooperate.

The only method to avoid this bias is to use a random method.

(c) Non-Sampling Errors

These errors arise from sources other than sampling. These errors include error of observation, errors of measurement and errors of responses. The physical procedures of observation can be imperfect which causes error. Errors in processing and analysis results in measurement errors. Incorrect responses of the respondents, mistakes, in noting their responses, etc. are called errors of response.

(d) Non-Sampling Biases

They consist of biases of observation and non-observation, response biases and process biases. When the observation are obtained and recorded incorrectly it causes biases of observation. Failure to obtain observations on some segments of the population due to either non-coverage or non-response results into non-observations biases. Response biases consist of biases arising from imperfections in field observations or interviewing. Processing biases are produced during coding, tabulation and computing.

Sample Design Characteristics -

Based on the factors explained earlier characteristics of the sample design are as follows Sample design must be -

(a) Representative

It should be representative to give a true picture of the population.

(b) Adequacy

The size of the sample should be adequate to provide reliability.

(c) Independence

All items of the sample should be selected independently of one another.

(d) Homogeneity

There should be no basic difference in the nature of units in the universe and in the sample

(e) Lack of bias

It should be unbiased and obtained by a random method or probability process.

(f) Smallness in size

The size of the sample should be small and economical.

(g) Accuracy and completeness

A sample should never give incomplete information and it should not omit any unit included in the sample.

Types of Sample Design

The sample design are classified on the basis of representation and selection technique. On the basis of representation it is classified as probability and non-probability, sample.

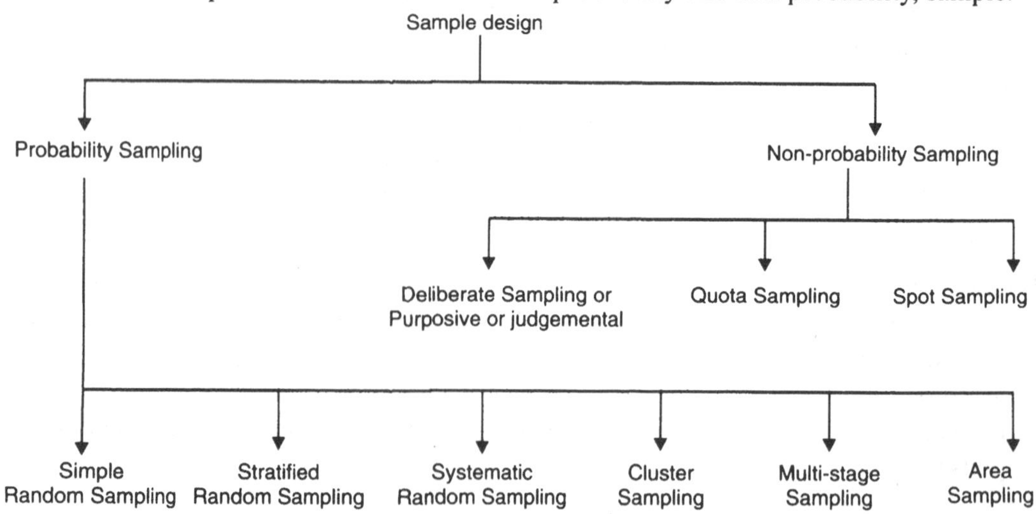

I Representative basis.

A. Probability sampling is based on the concept of random selection.

B. Non-probability sampling is non-random.

II Selection basis

Sample may be restricted or unrestricted.

Probability Sampling is of following types:

(1) Simple Random Sampling
(2) Stratified Random Sampling
(3) Systematic Random Sampling
(4) Cluster Sampling
(5) Area Sampling
(6) Multi-Stage Sampling
(7) Double Sampling.

Non-Probability Sampling may be classified into:

(1) Deliberate Sampling or Purposive Sampling or Judgmental Sampling.
(2) Quota Sampling.
(3) Spot Sampling.

Probability Sampling

It is known as random or chance sampling. In this system every item of population has every chance to be included. Its best technique for representative sample.

In random sampling there in equal probability of item to be included in sample. The actual practice for random sampling is done with the use of random number table.

Non-Probability Sampling

It is also known as deliberate purposive or judgement sampling. In this type of sample items are selected deliberately by researcher. There is no assurance that every element has some specifiable chance of being included.

The researcher may select sample favourable to research study and if researcher is unbias than there is relibility of the result. Quota sampling is also an example of non probabily sampling. All the various types of sampling are explained in the detailed.

1. Deliberate Sampling

This is also known as purposive, judgement sample. It invaulved the deliberate or convience selection of sample. It constitutes the representation of all universe. If data of petrol users is to be collected in relation to their reaction of quality of the petrol, than researcher may select few petrol pumps and conduct interview of the pet buyers. This is called convience sampling. In care of judgement sampling the judgement of the researcher is used for selecting sample.

2. Simple Random Sampling

This is also known as chance sampling. In this procedure of sampling design their are equal chances of the any item to get selected. The sample in this system are controlled by the probability Random number table used here, for selecting simple known as random sampling.

3. Systematic Sampling

This is most rational and practical way for selecting sample. In cause test of 100 person is available for slowly to we select so every 5th person i.e. 5, 10, 15 etc. is selected as systematic sample. But this type of sampling in possible only when such a rise in available for the reference.

4. Stratified Sampling

This method of selecting sample is used in selecting items when there is no homogenous group in the universe. So the universe is classified in different level of the strata to selected sample belonging to each strata on the basis of chance sampling. Therefore first classification is made and there chance sampling is done. Its known as stratified sampling. In the composition population in the region, if increasing demand is required to be assessed by increasing income than as there is no homogenous group so the population will be classified as Higher - Middle and Lower income group than based on chance sampling items will be selected from each of the strata of the classified group.

5. Quota Sampling

If the classification and chance selection as explained in stratified sampling in becoming costly, than these strata are subclassified in quota and selection of items in quota are left to the judgement of interviewer. The quota is normally in proportion to the strata. This excercise is condcted as non-probability sampling.

As we have seen classified strata Higher - Middle or lower income grap than HIG in 20%, M.I.G 30% and L.I.G. 50% in the compostion of population quota is proportionately worked cut as per the size of the strata 20%, 30%, 50% judgement sampling will be done in quote sampling in the ratio of 2:3:5.

6. Area Sampling

This sampling is done in relation to the geographical area under study. Total area is divided in the small region. These small geographical regions are selected randomly. All these small region are included in sample.

This sampling in possible only when information of area grouping is available. For the purpose of area sampling Nation is divided in state, state in division, district, taluka, block and village group on the bases of the records available.

7. Cluster Sampling

This sampling is just like area sampling. The classification of population in group is called cluster and sample in selected from each of cluster. Suppose Bank wants to issue ATM card to its credit card holders to about 1,000 custmors using credit card. 1000 card holders are to be divided in 20 groups, each group will hane. 50 eardhaldous if size is to kept 300 card holders so 15 cardholders will be selected in each group randomly.

This will make work of investigation easier and increase the efficiency in the field investigation.

8. Multi - Stage Sampling

This is an extenstion of area or group sampling. When national level big studies are conducted then nation is divided state further in division thus in district and finally at village block level. Now states in the nation divisions in the state, districts in the division and lastly village bolocks in district are selected on random bases. So the group classification of various sloges is described multi stage sampling.

9. Spot Sampling

When it is not possible to fixed sample of the size in advance because of the complexity of sample design then it is decided at the spot in ongoing process. The progress of the survey is closely monitered and on the basis of the information received as the work progress the appropriate design is adopted by assessment on the spot.

Double Sampling

It may be more convenient or economical to collect some information by sample and then use this information as the basis for selecting a subsample for further study. This procedure is called **double sampling, sequential sampling,** or **multiphase sampling.** It is usaully found with stratified and/or cluster designs. The calculation procedures are described in more advanced texts.

Double sampling can be illustrated by the dining club example. You might use a telephone survey or another inexpensive survey method to discover who would be interested in joining such a club and the degree of their interest. You might then stratify the interested respondents by degree of interest and subsample among them for intensive interviewing on expected consumption patterns, reactions to various services, and so on. Whether it is more desirable to gather such information by one-stage or two-stage sampling depends largely on the relative costs of the two methods.

Because of the wide range of sampling designs available, it is often difficult to select an approach that meets the needs of the research question and helps to contain the costs of the project. To help with these choices, exhibit may be used to compare the various advantages and disadvantages of probability sampling. Nonprobability sampling techniques are covered in the next section. They are used frequently and offer the researcher the benefit of low cost. However, they are not based on a theoretical framework and do not operate.

Comparison of Probability Sampling Designs

Type	Description	Advantages	Disadvantages
Sample Random Cost : High Use : Moderate	Each population element has an equal chance of being selected into the sample. Sample drawn using random number table/generator.	Easy to implement with automatic dialing (random-digit dialing) and with computerized voice response systems	Require a listing of population elements. Takes more time to implement. Uses larger sample sizes. Produces larger errors.
Systematic Cost : Moderate Use : Moderate	Selects an element of the population at the beginning with a random start, and following the sampling skip interval selects every kth element.	Simple to design. Easier to use than the simple random. Easy to determine sampling distribution of mean or proportion.	Periodicity within the population may skew the sample and results. If the population list has a monotonic trend, a biased estimate will result based on the start point.

Stratified Cost : High Use : Moderate	Divides population into subpopulations or strata and uses simple random on each stratum. Results may be weighted and combined.	Researcher controls sample size in strata. Increased statistical efficiency. Provides data to represent and analyze subgroups. Enables use of different methods in strata.	Increased error will result if subgroups are selected at different rates. Especially expensive if strata on the population have to be created.
Cluster Cost : Moderate Use : High	Population is divided into internally heterogeneous subgroups. Some are randomly selected for futher study.	Provides an unblased estimate of population parameters if properly done. Economically more efficient than simple random. Lowest cost per sample, especially with geographic clusters. Easy to do without a pupulation list.	Often lower statistical efficiency (more error) due to subgroups being homogeneous rather than heterogeneous.
Double (sequential or multiphase) Cost : Moderate Use : Moderate	Process includes collecting data from a sample using a previously defined technique. Based on the information found, a subsample is selected for further study.	May reduce costs if first stage results in enough data to stratify or cluster the population.	Increased costs if indiscriminately used.

Use of the Normal Distribution

The samples in surveys are often large enough so that an estimate made from them is approximately normally distributed. Further, with probability sampling, we have formulas that give the means and variance of the estimate. Consider first unbiased estimates. An estimate β given by a sampling plan is called an unbiased estimate of some population characteristic μ is the mean value of β, taken over all possible samples, is equal to μ. In the this condition may be written.

$$E(\mu) = \sum_{i=1}^{V} \pi_i \beta_i = \mu$$

where, β_i = is the estimate given by the th sample.

Σ = the expected value of is used frequently.

Suppose that we have taken a sample by a procedure known to give unbiased estimates and have computed the sample estimate β and its standard deviation σ_a (often called, alternatively, its standard error). How good is the estimate? We cannot know the exact value of the error of estimate $(\beta - \mu)$, but from the properties of the normal curve the chances are

0.32 (about 1 in 3) that the absolute error $\beta - \mu$ exceeds $\sigma\beta$

0.05 (1 in 20) that the absolute error $\beta - \sigma$ exceeds 1.96 $\sigma_a = 2\sigma\beta$

0.01 (1 in 100) that the absolute error $\beta - \sigma$ exceeds 2.58β

For example, if a probability sample of the records of batteries in routine use in a large factory shows an average life $\beta = 394$ days, with σ_a standard error a = 4.6 days, the chances are 99 in 100 that the average life in the population of batteries lies between

$\beta_L = 394 - (2.58)(4.6) = 382$ days.

and

$\beta_U = 394 + (2.58)(4.6) = 406$ days

The limits, 382 days and 406 days, are called lower and upper confidence limits. With a single estimate from a single survey, the statement "μ" lies between 382 and 406 days" is not certain to be correct. The "99% conficence" figure implies that if the same sampling plan were used many times in a population, a confidence statement being made from each sample, about 99% of these statements would be correct and 1% wrong. When sampling is being introduced into an operation in which complete censuses have previously been used, a demonstration of this property is sometimes made by drawing repeated samples of the type proposed from a population for which complete records exist, so that μ is known (see, e.g., Trueblood and Cyert, 1957). The practical verification that approximately the stated proportion of statements is correct does much to educate and reassure administrators about the nature of sampling. Similarly, when a single sample is taken from each of a series of different populations, about 95% confidence statements are correct.

The preceding discussion assumes that σ_a as computed from the sample, is known exactly, Actually, σ_a like β, is subject to a sampling error. With a normally distributed variable, tables of Student's *t* distribution are used instead of the normal tables to calculate confidence limits for μ when the sample by the *t* table makes almost no difference if the number of degrees of freedom in σ_a exceeds 60, With certain types of stratified sampling and with the method of replicated sampling the degrees of freedom are small and the *t* table is needed.

Selection of Random Sample -

The probability of drawing any one of the element in the first draw is 3/6 the probability of drawing one more element in the second draw in 2/5 and the third draw in 1/4 without replacement.

Joint Probability workout to $\frac{3}{6}, \frac{2}{5}, \frac{1}{4} = \frac{6}{120} = \frac{1}{20}$

Various statisticians like Tippet, Yates, Fishor have prepared tables of random numbers which can be used for selecting random sampling.

Random Sampling Design Systematic Sampling

If a 4% sample is desired the first item would be selected randomly from first 25 and there after every 25th item would be included.

Stratified Sampling :

The population is divided into several sub population that are individually more homogeneous and there we select items.

Suppose we want sample size of n = 15 to be drawn from the population of size = N = 4000.

Dividing into three strata of size

N1 = 2000, N2 = 1200 and N3	=	800
By adopting proportional allocation		
We shall get sample size as		
For strata with N1 we have =	2000	
P1 = 4000/8000 and hience	=	2000/4000
N1 = N P1 = 30 (4000/8000)	=	15 (2000/4000) = 7.5
For strata with N2 = , we have	=	1200
n2 = n P2 = 30 (2400/8000) = 09	=	15 (1200/4000) = 4.5
For strata with N3 = we have	=	800
n3 = n°P3 = 30 (1600/8000) = 06	=	15 (800/400) = 3.0

So the sample size for different strata are 7.5, 4.5 and 3 which is in proportion to the size of the strata.

Problem No. 1

A population is divided into three strata

$N_1 = 10{,}000$, $N_2 = 4000$ an $N_3 = 6000$.

Respective standard deviations are

$\delta_1 = 30, \ \delta_2 = 36, \ \delta_3 = 10,$

How sample size n = 168 to be alloted in three strata it optimum allocation using disproportionate sample design ?

Sollution

By applying disproportionate sample design for optimum allocation the sample size in respect of dillrect will be determined as follow :

Sample size for strata with $N_1 = 10{,}000$

$$N_1 = \frac{168\,(10{,}000)\,(30)}{(10{,}000)\,(30) + (4000)\,(36) + (6000)\,(10)}$$

$$= 5{,}04{,}00{,}000 \,/\, 5{,}04{,}000 = 100$$

Sample size for strata with $N_2 = 4000$

$$N_2 = \frac{168\,(4{,}000)\,(36)}{(10{,}000)\,(30) + (4000)\,(36) + (6000)\,(10)}$$

$$= 2{,}41{,}92{,}008 \,/\, 5{,}04{,}000 = 48$$

Sample Size for strata with $N_3 = 6000$

$$N_3 = \frac{168\,(6{,}000)\,(10)}{(10{,}000)\,(30) + (4000)\,(36) + (6000)\,(10)}$$

$$= 1{,}00{,}80{,}000 \,/\, 5{,}04{,}000 = 20$$

So 168 Sample size can be alloted in the strata as under

N_1 = 100
N_2 = 48
N_3 = 20
= 168

Factory outlets :

A	70	I	52	If we want to select a sample
B	34	J	132	of 10 outlets using cities as
C	20	K	74	cluster and selecting within
D	64	L	88	duster proportional to size,
E	140	M	66	how many outlets from each
F	56	N	58	city should be selected
G	52	O	56	(using 20 as starting point)
H	38			

City	No. of outlets	Cumulative Total	Sample	
A	70	70	20	The number of actual
B	34	104		from the city as given
C	20	124	120	in the sample to
D	64	188		select 10 randomly
E	140	328	220	
F	56	384		
G	52	436	420	2 stores from cdy (8)
H	38	474		1 store from 8 calls
I	52	526	520	from A, B, C and
J	132	658	620	I, J, K, L, N
K	74	732	720	

L	88	820	820
M	66	886	
N	58	944	920
O	56	1000	

In this example 1000 outlets are there and to select 10 outlets so the 1000 outlets/ 100 stores, so the sampling interval is 100. As starting point is 20 we will keep on odding 100 till co is selected

Problem No. 2

The following are the number of chain shops in 10 cities 70, 54, 48, 64, 84, 60, 68, 88, 58, 76. If we have to select is stroes using cities as duster and selecting within clusters proportional to size how many stros from each city should be select (using starting point as 8)

City	Stores No.	Cumulative Number	Sample
1	70	70	8
2	54	124	52 - 96
3	48	172	140
4	64	236	184-228
5	84	320	272-316
6	60	380	360
7	68	448	404-448
8	80	528	492
9	58	586	536-580
10	76	662	624

Total stores are 662 to select 10 sample stres the sampling interview 662/ 1015 44 starting poing 8 So we will doing increment to 8 by 44 till 15 stras are selected.

15 stores will be selected as under

2 from 2, 4, 5, 79	=	5 (5 x 2)	=	10
1 from 1, 3, 6, 8, 10	=	(5 x 1)	=	05
				15

QUESTIONS

1. What are the steps involved in Sample Design and explain various types of sample design.
2. What do you mean by 'Sample Design'? What points should be taken into consideration by a researcher in developoing a sample design for his research project.
3. Under what circumstances stratified random sampling design is considered appropriate? How would you select such sample? Explain by means of an example.
4. Explain and illustrate the procedure of selecting a random sample.
5. Compare probability sampling designs.
6. Explain your ideas about double sampling.

■■■

CHAPTER 4

METHODS OF DATA COLLECTION

After going through this chapter, you will be able to know :

- Sources of data
- Methods of data collection
- Primary data
- Observation method
- Limitation of observation method
- Classifciation of observation method
- Interview Method
- Definition of Interviews
- Objectives Interview
- Types of Interview
- Interviewer Qualifications
- Survey through questionnaries and schedules
- Data collection through Questionnaire
- Definition of Questionnaire
- Computer direct interviews

The collection of data is the important task in the research methodology. Before explaining the methods of data collection, researcher, should understand the need of the study and decide the type of the data required.

Sources of Data

It is necessary to distinguish between types of the data. It is primary and secondary one. The primary date is collected at first hand and the secondary data are those which is collected by some one else and which has already been processed.

The distinction between primary and secondary data can be made more clear on the basis of documents. The primary data is documented as record and secondary data as report.

Now researcher has to decide which type of data he would like to use for this study and accordingly he will have to select particular type of data base. The data collection methods differ in each type of data to be collected by the researcher personally, where as in secondary data it is only compilation of the data already collected.

Methods of data collection

Primary Data

The primary data is collected through several methods. The important among then are

A. Observation

B. Interview

C. Survey through questionnaires and schedules.

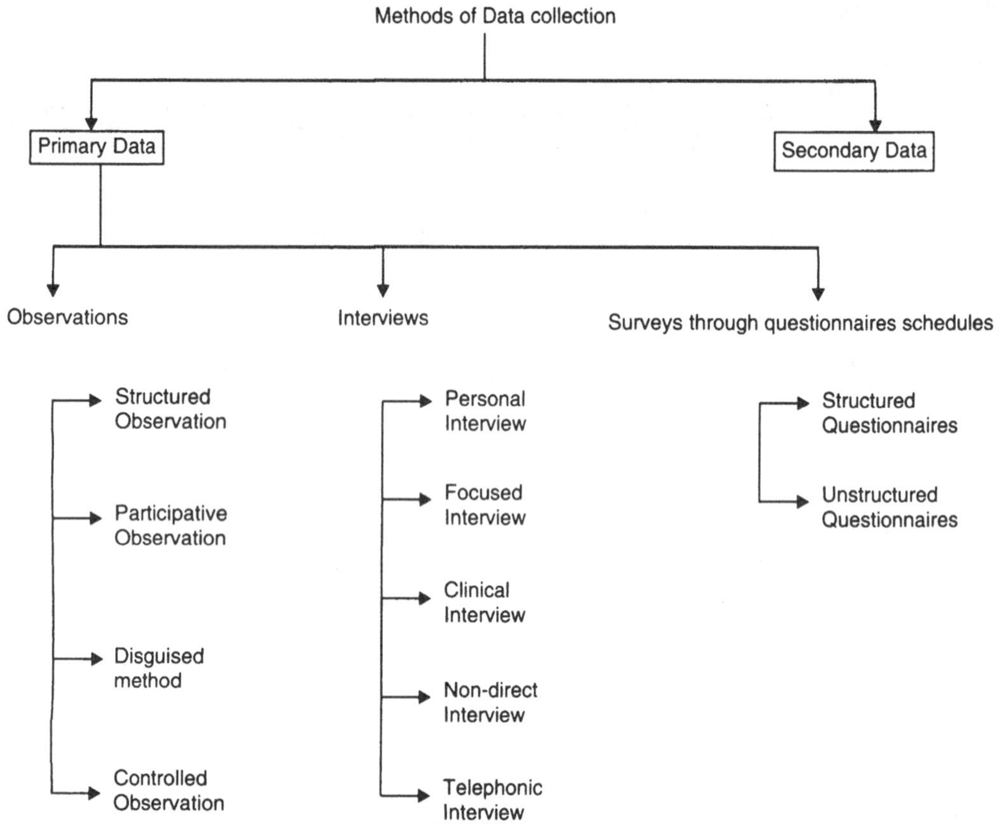

A. Observation Method

Observations may take place in laboratory or in real life situations. This method is commonly used in the study of behavioral sciences. In this method the observer may participated in the-group observed by him or he can do the observation from outside the group and as such his presence may be unknown to the people, so they act freely.

The observation is systematically plan and it is considered as scientific tool for primary data collection. The reliability at this data can also be checked.

In observation method information is collected on the basis of observation carried out without asking the respondent.

Advantages of Observation Method

1 Subjective bias is eliminated if accurate observation is made
2. It is not affected by past behaviour for future attitudes.
3. Active participation of respondent is not required.

Limitation of Observation Methods

1. It is an expensive method.
2. The information provided in this method is very limited.
3. Some unforeseen factors may affect the work of observation.
4. If respondent is not accessible than observation losses its effectively :

Classification of Observation

The observation may be broadly classified as under

(a) Structured Observation

When observation is made by careful definition, recording of observation is standardised, it is called structured observation. It is appropriate in descriptive studies.
If the observation is not made on well thought conditions it is called unstructured observation. This unstructured observation is conducted in exploratory study.

(b) Participative Observation

In this method of observation, observer may make himself as member of group sharing the experience with the group, it is called participant observation.
But when he is detached from group for direct participation in group and he will have no experience about feeling of members of the group, this type of observation is known as non-participant observation.

Merits of Participative Observation

1. The natural behaviour of group is recorded.
2. Such information is collected which can not be collected in case of non-participative method.
3. Truth of information can be assessed.

Demerits of Participative observation method :

1 The observation may lose the objectivity with emotional participation with group.
2 In this method the observasation experience may narrow down.

(c) Disguised method

In this method observation is made in such a way that his presence is not known to the members of the group.

(d) Controlled Observation

When observation is conducted with pre decided plans with certain procedures it is termed as controlled observation.

In controlled observation mechanical instruments are used for standardization and accuracy. It takes place in laboratory in controlled conditions.

Uncontrolled observations are those which are conducted in natural conditions. No instrument is use to assertion the accuracy. This observation is made to know real life situations so more time is required to be allotted for this observation. This method is used in exploratory research study.

The use of observation method depends upon the following factors -

1. Researchers biases and his prejudices,
2. 'His power of perception, outlook and ability to recognise casual relationship and accuracy in recording results of his observation

B. Interview Method

The interview method is one of the important methods of primary data collection. It is a confiscation between the observation and respondent. It is oral-verbal questions and corresponding oral - verbal response to the quaries made. In field oriented study interview plays a major role.This is done with the help of structured schedule. The interview can be conducted personally or though telephones.

As a research method, we can regard interviewing as the development of the common social act of conversation that is characterized by the concepts of precision, focus, reliability, and validity. The interviewer asks questions pertinent to the research and the respondent answers the questions. The concept of an interview, usually understood as a face-to-face encounter, can be extended to include telephone interviews and, in today's context, video interviews.

Interviews can be regarded as one of the foremost tools of the behavioural scientist, not only for the purpose of obtaining qualitative data but also as a tool for measurement. When seeking measurement, the questions asked in on interview are so expressed as to elicit responses that can be recorded in categories and coded. Nevertheless, interviews are particularly valuable for obtaining data in situations where the formality of a questionnaire would be less likely to elicit the desired information. Thus an interview can be used as an exploratory tool to suggest hypotheses and identify research variables as well as for studying relationships and testing hypotheses.

Definition of Interviews

As a method of scientific social research, several scholars have tried to define interview.

P. V. Young says, "The Interview may be regarded as a systematic method by which one persons enters more or less legitimately into the inner life of another who is generally a stranger to him.

Hsin Pao Yang observes, "The interview is a technique of field work which is used to watch the behaviour of an individual or individuals, to record statements, to observe the concrete results of social or group interactions.

While defining Interview.

CA. Master holds that "in a formal interview pre-determined questions are asked and the answers are collected in a certain way.

In simple words, we can say that interview is it method of data collection in which the researcher meets the respondents and collects information about his study area through talks with them or by putting Tormal questions to them and by recording their answers.

Objectives of Interview

1. The primary objective of interview is to collect information about unknown facts through personal contact. For instance a histrocial researcher wanting to write about the contribution of those from a particular region who participated in the freedom struggle, may have to resort to only interview such people, their friends and relatives to get the necessary details.

2. Collection of qualitative facts can be done only through interview.

3. The interview method is adopted to suppliment the data and inferences obtained through observational technique.

4. Collection of information about different problem in varied situations.

5. Interview technique helps to get facts in the formulation of hypothesis.

Types of Interviews

Interviews broadly categories as under

(i) Personal Interview

It is face to face interaction with observer and respondent. In this method the interviewer ask various questions and collects necessary information. This is a direct personal investigation from the source of information.

The observer or interviewer has to present on the spot to meet the person from whom this information is to be collected.

If direct personal contact is not possible for observer than indirect oral examination can be conducted which should be subject to cross examination by the person Conducting investigation.

The interview is normally carried out in structured way. Such interviews are called structured interviews. This is conducted through well thought pre-determined questions with sound technies of recording the response. It is a rigid procedure for asking questions.

But in case of unstructured interview there is no rigidity in asking question maximum flexibility is mentained in questioning. As such in such interview no pre-determined set of questions are used for collecting information. In this system interviewer is given greater freedom to decide questions if required additional questions can also be asked.

On account of having such flexibility in questions the comparesion between two interviews is very difficult. It requires deep knowledge and greater skill for collecting information. This unstructured interview is suitable for formulative or explootory research studies. Where as structured interview are used for descriptive studies. It is economical, used for generalisation and required normal skill to conduct interview.

(ii) Focused Interview

When experience of respondent is given along with its effects it is called focused interview. In this interview observer has freedom to decide the sequence of the questions.

The respondent is confined to the issues considered conversance. These interview are generally used in developing hypothesis and conducted mostly as unstructured one.

(iii) Clinical Interview

This interview is mainly concerned with individual feelings motivation and experience. Under this method of data collection information is left to the discretion of the interviewer.

(iv) Non-direct Interview

In this method in place of direct interview the respondent is encouraged to express his opinion on the topic given to him for discussion. As an observer interviewer assess the feeling and beliefs of the respondent.

Advantages and disadvantages of interview method

The various methods of interviews are explained above, which has advantages and disadvantages which are enumerated as under.

Advantages of Interview method

1. In-depth information is obtained.
2. Interviewer can overcome the resistance by respondent.
3. Questions can be restructured as per the requirement.
4. By observation verbal answers to can be recorded.
5. Under this method personal information is easily obtained.
6. Samples can be easily controlled.
7. Non-responsence by the respondent is minimised.
8. Spontaneous response of the Yespondent is secured.
9. Language of the_ questions can be modified according to the level and understanding of respondent.

10. Based on the response of the respondent if required additional questions can be asked by interviewer.
11. It helps to obtain accurate information.
12. Its a process of social interaction.
13. Misunderstanding of questions can be removed by better clarification by interviewer.
14. Personal qualities of interviewee, such as emotions opinions are assessed by interviewer.

Disadvantages of Interview Method

1. This method in very expensive where large sample is taken.
2. There is possibility of bias interviewer and interviewee, hence more supervision is required.
3. The data so collected may remain inadequate if high profile persons are not contactable.
4. This process is more time consuming.
5. Over response by respondent may lead to imaginary answer.
6. Under this method training and supervision is required to be organised which is very complex.
7. Under the process of interviewing systematic errors may introduced.

Pre-Conditions for Interview Method

1. Careful selection of Interviewer and proper instructions to him.
2. Interviewer must possess technical competance must have personal qualities like Honesty, sincerety and impartiality.
3. Continious supervision and field checks are necessary.
4. Efforts should be made to crate atmosphere of trust and confidence. Irrelevant conversation should be discouraged.

(v) Telephonic Interview

When respondent, can not be contacted personally then information which is required to be collected, can be collected through phone. This method is not widely used but even than it is a main source for survey in developed regions.

Advantages of this system

1. This system is having more flexibility.
2. It is quick way for contacting respondent.
3. It has relatively low cost, call back is very easy.
4. Non-response is very low.
5. There is no embarrassment to respondents.
6. Interview requirement can be easily explain, to respondent.
7. Easy access to respondent who are otherwise uncontactable.
8. Wider range of the sample is possible under this system.

Disadvantages

1. Very short time is given to respondent for his answers.
2. This system is useful only in such place where, network is developed.
3. Wide coverage becomes costly.
4. It is difficult to have descriptive answer to the questions
5. Higher possibility of bias interview is more.

6. Short questions may lose significance.
7. There is much scope for deckling answers by respondents.
8. This system lacks motivation as there is no eye contact.

Interviewer Qualifications

Interviewing requires a trained interviewer or the skills gained from experience. These skills include making respondents comfortable, probing for detail without making the respondent feel harassed, remaining neutral while encouraging the participant to talk openly, listening carefully, following a participant's train of thought, and extracting insights from hours of detailed descriptive dialogue. Skilled interviewers learn to use their personal similarities with or differences from their interviewee to mine for information; similarities are used to convey sympathy and understanding, while differences are used to demonstrate eagerness to understand and empathize.

In quantitative research we are more interested in the data collector's following a prescribed procedure. Whereas in qualitative research the individual conducting the interview needs a fuller understanding of the dilemma and how the insights will be used. So a skilled interviewer must be a "quick-study," someone who can grasp an understanding of an issue without necessarily having prior experience with the product or service or being a technical expert.

The interviewer needs to be able to extract information from a willing participant who often is not consciously aware that he or she possesses the information desired. The actual interviewer is usually responsible for generating the interview or discussion guide, the list of topics to be discussed (unstructured interview) or the questions to be asked (semistructured) and in what order (structured). In building this guide, many interviewers employ a hierarchical questioning structure, depicted in Exhibit the following. Broader questions start the interview, designed to put participants at ease and give them a sense that they have a lot to contribute, followed by increasingly more specific questions to draw out detail.

The interview Question Hierarchy

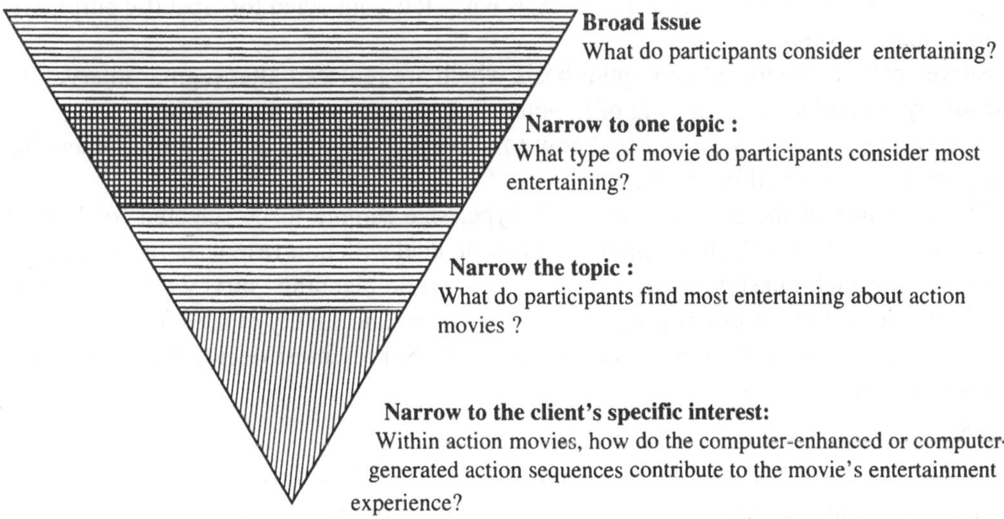

The interviewer is often responsible for generating the screening questions used to recruit participants for the qualitative research. This preinterview uses a device similar to a questionnaire, called a recruitment screener. The above diagram provides the various elements necessary for a comprehensive recruitment screener. Each question is designed to reassure the researcher that the person who has the necessary information and experiences, as well as the social and language skills to relate the desired information, is invited to participate. Data gathered during the recruitment process are incorporated into the data analysis phase of the research, as recruitment data provide additional context for participants' expressions.

In general, then, the interviewer is a consultant with wide-ranging responsibilities.

- Recommends the topics and questions.
- Controls the interview, but also plans - and may manage - the locations and facilities for the study.
- Proposes the criteria for drawing the sample participants.
- Writes the recruitment screener and may recruit participants.
- Develops the various pretasking exercises.
- Prepares any research tools (e.g., picture sorts or written exercises) to be used during the interview.
- Supervises the transcription process.
- Helps analyze the data and draw insights.
- Writes or directs the writing of the client report, including extracting video clips for the oral report.

C. Survey through Questionnaries and Schedules

Survey has been defined as a fact finding exercise for discovery of facts. The object of the survey is to explain and provide information for formulation of hypothesis.

The surveys broadly conducted for demographic characteristics of set of people for opinion and attitudes, social activities and environment.

When surveys are conducted for collecting general information without any specific emphasis it is know as general survey where as when it is conducted for Specific purpose it is called specific survey.

Survey can be conducted on regular basis which are repeated after regular interval. In case survey is conducted for testing of hypothesis it is called ad hock survey.

The purpose of survey should be know to surveyors for efficient enquiry. Otherwise quality of the survey will be affected.

The coverage of the survey will decide type, size sample to be selected for having desired accuracy. It should also explain as to how to deal with non-response. In survey data is collected through questionnaries and schedules, while planning survey errors must be anticipated along with necessary precautions to be taken. The quality of the survey is largely depends upon training and supervision of field staff. Survey must decide time period for completion of work.

Pilot Survey

It' is Survey on small scale. It provides the information on the following aspects.

1. It enables the researcher to have proper knowledge of the universe to be sampled.
2. How non response can be eliminated.

3. The adequacy of the sample design for survey.
4. The suitability of alternative method of data collection is decided from pilot survey.
5. It also test the efficiency of briefing provided to staff.
6. It also provide the probabletest of the main survey to be conducted.

Advantages of Survey Method

1. This enables the researcher to establish contact with respondent.
2. By survey there is scope for introducing new theory.
3. It facilitates the researcher to have observation of emotional impact on the situation.
4. It provides information to practical problems.
5. It is conducted to investigate cause and effect relationship of the problem under study.

Disadvantages of Survey Method

1. As surveys are conducted on sample basis the respondent may not be proper representative of the population.
2. It is for study of current problems only.
3. Long term development study is not possible in this method.
4. It requires huge funds to take-up training, supervision, tabulation and analysis of data.
5. The reliability,and validity is questionable if surveys are conducted in hurry.
6. If survey is not perfect and pre-plan. It will not give any desired result.

Data Collection through Questionnaires

In case of large surveys to be conducted questionnaires are very popular method for data collection. It is adapted by all individuals and institution's.

It consist of number of questions in definite order in set forms. It can be mailed to respondent. They are expected to read and understand questions and answers given questions. The respondent are requested to mail back the dully filled in questionaries to the researchers.

Definition of Questionnaire

Questionnaire is a set of questions, a systematic list of questions which is sent to the respondents who answer these questions and return the questionnaire to the researcher.

Prof Hsin Pao Yang says, "In its simplest form, the questionnaire consists of a schedule of questions sent by mail to the persons on a list or in a sample survey." The persons to whom the Questionnaires is sent are called respondents, who are to send their answers to researcher.

Bogardus remarks that Questionnaire is a list of questions handed over to different persons who are to answer the questions.

From these two definitions of Questionnaire we can mention the following features of the Questionnaire method of data collection in the research.

1. Questionnaire is a list or a set of questions.
2. Questionnaire is prepared by the researcher. It contains questions relating to the topic of his research.
3. The purpose of each question of the Questionnaire is to elicit specific information from the respondents.
4. The questionnaire is sent to the respondents and they send their answers back to the researcher.

5. Only educated and willing respondnets can take part in this exercise.
6. In questionnaire method there is no direct contact between the researcher and the respondents.
7. Questionnaire method of data collection is very useful in studying a geographically scattered universe of research.
8. Since there is no direct contact between the researcher and the respondents, there is no danger of any influence of the former on the latter.
9. Data collected through Questionnaire method has a high level of accuracy.
10. Questionnaire method is a time and money saving method.
11. It is a direct method of Data Collection.
12. Questionnaire method is a recognised scientific method of social research.

i. Mail Survey

The method of collecting data though mailling of questionnaries to respondent is called mail survey. This system is widely used in various business and economic survey.

1. Inspite of large area to be covered it is low cost.
2. Respondent answer the question in his own word get answer the questions.
3. Respondent can be conveniently contented.
4. On account of large sample, the results are more reliable.

In mail survey

1. Non-response may be high/, Bch result low rate of duly filled in forms.
2. Utility of the system is largely based on co-operation extended by the respondent.
3. No flexibility in questionnaries once they are sent to respondent.
4. Possibility of omission and ambiguous replies is always there.
5. This is time consuming method.

If questionnaries maillin is becoming difficult, can not be used than this job is required to be done by investigator. They will personally contract respondent, request them for filling of questionnaries - i ll required to collect the dully filled in questionaries. The problems in mailling can be eliminated in personal contact but it will be very expensive. Investigator may required to visit the again and again in case of non-response.

Advantages of Mail Surveys

- Mail surveys are among the least expensive.
  This is the only kind of survey you can do if you have the names and the target population, but not their telephone numbers.
  The questionnaire can include pictures - something that is not possible phone.
- Mail surveys allow the respondent to answer at their leisure, rather than often inconvenient moment when they are contacted on phone in person. For this reason, they are not considered as intrusive as other kinds of interview.

Disadvantages

- Mail surveys take longer time than other kind of survey.
- You will need to wait several weeks after mailing out questionnaires before you can be sure that you have for gotion most of the responses.

- In populations of lower educational and literacy levels, response rates mail surveys are often too small to be useful. This, in effect, eliminates many immigrant populations that form substantial markets in many areas. Even in well educated populations, response rates vary from as low as 3% up to 90%. As a rule of thumb, the best response levels are achieved from highly-educated people and people with a particular interest in the subject (which, depending on your target population, could lead to a biased sample).

One way of improving response rates to mail surveys is to mail a postcard telling your sample to watch for a questionnairc in the next week or two. Another is to follow up a questionnaire mailing after a couple of weeks with a card asking people to return questionnaire. The downside is that this doubles or triples your mailing cost. If you purchased a mailing list from a supplier, you may also have to pay a second and 3rd use fee - you often cannot buy the list once and re-use it.

Another way to increase responses to mail surveys is to use an incentive. One possibility is to send a money consideration (or more) along with the survey. Many people will consider their time worth more than a money consideration. Another possibility is to inculde the people who return completed surveys in a drawing for a prize. A third is to offer a copy of the (non-confidential) result highlights to those who complete the questionnaire. Any of these teachiques will increase the response rates.

Remember lhat if we want a sample of 1,000 people, and we estimate a 10% response level, we need to mail 10,000 questionnaires. We May want to check with your loacal post office about bulk mail rates - you can save on postage using this mailing methods However, most researchers do not use bulk mail, because many people associate bulk with "junk" and will throw it out without opening the envelope, lowering our response rate. Also bulk mail moves slowly, increasing the time needed to complete tour project.

Computer Direct Interviews

These are interviews in which the Interviewees enter their own answers directly in to a computer. They can be used at trade shows, offices, and so on. The survey System's optional Interviewing Module and Interview Stations can easily creative computer-direct interviews. Some researchers set up a Web page survey for this purpose.

Advantages

- The virtual elimination of data entry and editing costs.
- You will get more accurate answers to sensitive questions. Recent studies of potential blood donors have shown respondents were more likely reveal HIV related risk factors to a computer screen than to either human interviewers or paper questionnaires. It was found that computer-aided surveys among drug users get better results than personal interviews. Employees are also more often willing to give more honest answers to computer than to a person or paper questionnaire.
- The elimination of interviewer bias. Different interviewers can ask questions in different ways, leading to different results. The computer asks the questions same way every time.
- Ensuring skip patterns are accurately followed. The Survey System can ensure people are not asked questions they should skip based on their earlier answers. These automatic skips are more accurate than relying on an Interviewer reading paper questionnaire.

- Response rates are usually higher. Computer-aided interviewing is still novel enough that some people will answer a computer interview when they would not have completed another kind of interview.

Disadvantages

- The interviewees must have access to a computer or one must be provided for them.
- As with mail surveys, computer direct interviews may have serious response rate problems in populations of lower educational and literacy levels. This method may grow in importance as computer use increases.

E-mail Surveys and Web Page Surveys

Email surveys are both very economical and very fast. More people have email than have full Internet access. This makes email a better choice than a Web page survey for some populations. On the other hand, email surveys are limited to simple questionarries whereas Web page surveys can include complex logic.

Advantages of E-mail Surveys

- Speed. An email questionnaire can gather several thousand responses within day or two.
- There is practically no cost involved once the set up has been completed
- You can attach pictures and sound files.
- The novelty element of an email survey often stimulates higher response level than ordinary mail surveys.

Disadvantages of E-mail Surveys

- We must possess (or purchase) a list of email addresses.
- Some people will respond several times or pass questionnaries along to friends to answer. Many programs have no check to eliminate people responding multiple times to bias the results. The Survey System's Email Module will only accept one reply from each address sent the questionnaire. It eliminates duplicate and pass along questionnaires and checks to ensure that respondents have not ignored instructions (e.g:, giving 2 answers to a question requesting only one.
- Many people dislike unsolicited email even more than unsolicited regular mail you may want to send email questionnaires only to people who expect to get e-mail from you.
- You cannot use email surveys to generalize findings to the whole populations. People who have email are different from those who do not, even when matched on demographic characteristics, such as age and gender.
- Email surveys cannot automatically skip questions or randomize question or answer choice order or use other automatic techniques that can enhance survey the way Web page surveys can.

Many email programs are limited to plain questionnaires and can not show pictures. Email questionnaires from The Survey System can attach graphic or second file. Although use of email is growing very rapidly, many "average" citizens still do not possess email facilities, especially older people those in lower income and education groups. So email surveys do not reflect the population as a whole. At this stage they are probably best used in a corporate environment where email is common or when most members of the target population are known to have email. Internet/Internet (Web Page) Surveys Web surveys are rapidly gaining popularity. They have major speed, cost and flexibility advantages, but also significant sampling limitations. These limitations make software selection especially important and restrict the groups you can study using this technique.

Advantages of Web Page Surveys

- Web page surveys are extremely fast. A questionnaire posted on a popular. Web site can gather several thousand responses within a few hours. Many people who will respond to an email invitation to take a Web survey will do so the first day and most will do so within a few days.
- There is practically no cost involved once the set up has been completed. Large samples do not cost more than smaller ones (except for any cost to acquire the sample).
- You can show pictures. Some Web survey software can also show video and play sound.
- Web page questionnaires can use complex question skipping logic, randomizations and other features not possible with paper questionnaires or most email surveys. These features can assure better data.
- Web page questionnaires can use colors, fonts and other formatting options not possible in most email surveys.
- A significant number of people will give more honest answers to questions about sensitive topics, such as drug use when giving their answers to a computer instead of to a person or on paper.
- On average, people give longer answers to open-ended questions on Web-page questionnaires than they do on other kinds of self-administered surveys.
- Some Web survey software, such as The Survey System, can combine the survey answers with pre-existing information you have about individuals taking a survey.

Disadvantages of Web Page Surveys

- Current use of the Internet is far from universal. Internet surveys do not reflect the population as a whole. This is true even if a sample of Internet users is selected to match the general population in terms of age, gender and other demographics.
- People can easily quit in the middle of a questionnaire. They are not likely to complete a long questionnaire on the Web as they would be if talking with a good interviewer. If your survey pops up on a web page, you often have no control over who -anyone from India or outside cruiising that web page may answer.
- Depending on your software, there is often no control over people responding multiple times to bias the results.

At this stage we recommend using the Internet for surveys mainly when your target population consists entirely or almost entirely of Internet users. Business-to business research and employee attitude surveys can often meet this requirement. Survey of the general population usually will not.

Another reason to use a Web page survey is when you want to show video or both sound and graphics. A Web page survey may be the only practical way to have many people view and react to a video.

In any case, be sure your survey software prevents people from completing more than one questionnaire. You may also want to restrict access by requiring a password (good software allows this option) or by putting the survey on a page that can only be accessed directoy (i.e., there are no links to it from other pages).

Contents of Questionnaire

The questionnaires should be carefully designed as it is considered as main instrument of the survey method. It facilities data collection. It is a verbal report for data by respondent. This will help researcher to get material which respondent is willing to share.

Advantages

1. It is more economical as compare to other method.
2. By well designed questionnaire the investigator will not required to exert more.
3. It is an impersonal technique.
4. Uniformity is provided by standardization of questions.
5. It ensures anonymity. The respondent has confidence they will not be identified For expressing' particular opinion.
6. It places less pressure on respondent for immediate response.

Disadvantages

1 It is useful only when respondent are literate.
2. In absence of supervisor monitoring return of duly filled in questionnaire is low.
3. If respondent misinterprets the question his reply will be mostly wrong, which is difficult to correct.
4. The utility of questionnaires is restrided to respondnet having clear views on the subject matter.
5. The success of the questionnaires depends upon the sense of responsibility of respondent.
6. It is not possible for researcher to vary the atmosphere having impact on subject according to his design.

Types of Questionnaires

It can be classified on various bases. The classification of questionnaires based on the variable of structure is as under

1 Structured questionnaires-
2. Unstructured questionnaires

1. Structured questionnaires - are those in which the questions are definite. Additional questions are restricted only in case of more detailed response. The questions are presented in same wording and in same order for having proper response.

The form of question may be either closed or open it means categorical or free. The question is set in advance.

In closed question fixed alternatives answers are given so respondent can choose appropriate one. In this question the response is limited to stated alternatives. These alternatives may be in form of Yes or No.

Open end questions are designed to allow free response on the subject where the issue is identified. The respondent in given the apportunity to answer the question on his own reference.

Advantages and Disadvantages

Closed Questions Advantages

1 Closed questions are simple to administer, quick and inexpensive to analyse. They are more efficient.

2. Alternatives replies helps to understand the clear meaning of the question.
3. Closed question requires respondent to make judgement about the alternatives

Disadvantages

(a) It may force statement of opinion on the issue.
(b) Many respondent may not have clear opinion on issue.
(c) Closed questions are ill-equiped to reveal issue.
(d) There question do not provide for respondent frame of reference.

Open Questions Advantages

1. It is used as an exploratory tool before opinion research objectives has been clearly defined.
2. Opinion of researcher is not force on respondent through suggestive answers.
3. In complex issue open ended questions are desirable.

Disadvantages

a. Different respondent are likely to make different interpretations quite different from researcher.
b. There is possibility that some of the questions may go undetected.

Unstructured Questionnaire

When the characteristics as explained above are not present in the questionnaire it is know as unstructured on non-structured questionnaire. Where pre decided questions can not serve the purpose of our research study than researcher is provided freedom for farming on the spot questions to get appropriate response. Mostly in this case of interview on the sport questions are asked.

In this process interviewer is provided with the guidelines about information is to obtained, than based on scope provided he will have to frame the question , and replies to this can be recorded by him. When research problem is being first explored the unstructured questions have more effectively.

Classification of Questionnaire

The information required in questionnaire may be classified in main three aspects.

A. Identification of Information.

Under this part of questionnaire researcher must identify the name and nature of survey conducted, name of the sponsoring agency object of survey, personal information of the respondent and statement in respect of confidential treatment to the information provided.

B. Background of Respondent

In this part of questionnaire socio-economic and political background information of respondent is collected. Age, Income, Social composition, political preference socio economic status items are included here.

C. Subject Matter of Survey

In this section questions will be asked about the understanding. When the knowledge of topic is clear than opinion on various issues, respondent will be holding, will be ascertained.

Incase if respondent advice is considered important than information will be collected accordingly. Sufficient information will also be provided to respondent to enable him to make proper judgment.

Schedule and Questionnaire

Schedule is also a set of questions which is filled in by investigator who is made responsible to collect information. The investigator approach to the respondent and record the responses of the respondent. In some cases respondent is encourage to record the answers to the questions with the help of investigator.

The success of this method is based on selections and training of investigator, to collect information. Investigator must have competency for cross examination if necessary, investors personal qualities like honesty hard working will improve the quality of the work. This system is very Useful for extensive enquiries. It is very expensive still normally adapted by state and large organization. Census survey is conducted by this method.

Distinction between Schedule and questionnaire

Schedule	Questionnaire
1. Filled in by investigator	1. Filled in by respondent.
2 More expensive	2. Less expensive.
3. Low non-response	3. High non-response
4. Identity of respondent is know	4. Identity of respondent is not clear.
5. Process is fast as the information collected and filled in by investigator.	5. Process is slow as respondent do not return questionnaire immediately.
6. In schedule information is collected from illiterate respondent_	6. A useful only where respondent is Co-operative and literate.
7. Coverage is restricted to visit of investigator	7. Wide converge of investigation is possible.
8. More accurate information can be collected.	8. Risk of collecting wrong information
9. Success related to honesty and competence of investigator	9. Success of this method relates to quality of questionnaire.
10. Physical appearance will not affect the response.	10. Good physical appearance will have better response.

Scanning Questionnaires

Scanning questionnaires is a method of data collection that can be used with paper questionnaires that have been administered in face-to-face interviews; mail surveys completed by an Interviewer over the telephone. The Survey System can produce paper questionnaires that can be scanned.. If they do not understand a question. Intimate, personal questions are sometimes best handled by mail or computer, where anonymity is most assured.

While The Survey System will let you easily combine surveys gathered using different mediums; it is not usually recommended that you do so. A mail survey will often not give the same answers as the same survey done by phone or in person. If you used one method in the past and need to compare results, stick to that method, unless there is a compelling reason to change.

Keep questionnaire short and simple. Avoid the temptation - to add a few more questions just because you are doing a questionnaire.

Place your questions into three groups must know, useful to know and nice to know. Allow a "Don't know" or "Not Applicable" response to all questions, except to those in which you are certain that all respondents will have a clear answer. In most cases those are wasted answers as far as the researcher is concerned, but are necessary alternatives to avoid frustrated respondents.

Sometimes "Don't Know" or "Not Applicable" will really represent some respondents" most honest answers to some of your questions. Respondents who feel they are being coerced into giving an answer they do not want to give often do not complete the questionnaire. For example, many people will abandon a questionnaire that asks them to specify their income, without offering a "decline to state choice. For the same reason, include "Other" or "None" whenever either of these is a logically possible answer. When the answer choices are a list of possible opinions, preferences or behaviors, you should usually allow these answers. Include Do'nt "Not Applicable," "Other" or "None" in a list of choices being read over the telephone or in person, but you should allow the interviewer the ability to accept them when given by respondents.

Advantage of Scanning

- Scanning can be the fastest method of data entry for paper questionnaires.
- Scanning is more accurate than a person in reading a properly completed questionnaire.

Disadvantages of Scanning

- Scanning is best suited to “Check the box” type surveys and bar codes. Scanning programs have various methods to deal with text responses, but all require additional data entry time.
- Scanning is less forgiving (accurate) than a person in reading a poorly marked questionnaire.

Designing of Questionnaire/Schedule

The following consideration should be borne in mind by researcher while designing questionnaire. Proper selection and phrasing of question will assure good number of returns, which ultimately leads to accuracy in findings. Therefore due consideration is also required for physical design and questionnaire.

1. The Co-operation of respondent is largely based on good physical appearance of questionnaire. A poor or unattractive questionnaire may cause low response.
2. The questionnaire or scheduled designing will also have consideration to respondent. Respondent of different level and well trained investigator to record information so their ability to record response is main consideration of designing.
3. The choice of wooding result in understanding of question. The informant should understand the spirit of the question.
4. Special attention should given of sequence of the question.
5. The purpose of question has to considered while designing. If purpose is to knowledge ability or attitude for informant than it should be design accordingly.
6. If period survey is to conducted than in order to have uniformly questionnaire should be designed.

7. Analysis is main consideration in designing, the question should be design in such a manner where the classification and tabulation can be mode easily.

Physical Appearance of Questionnaire/Schedule

1. Size - The size of questionnaire and schedule depends upon the scope and items to be included. Large size in suitable for mail survey questionnaire where as small size is recommend for schedule. Therefore size should not be cumbersome for respondent.
2. Poper quality and colour - The durability is one of the requirements of preserving records. Light colour and smooth surface is desirable for paper to be used for questionnaire or schedule.
3. Items arrangement in questionnaire - Questions should be properly arranged. If the question is depended on answer of earlier one than it should be given proper place.

Choice of Questions

1. The questions having direct bearing on problem of research study should be included in set of questions.
2. The question whose answers can not have effectively or accuracy should be avoided.
3. The question should be included with an eye on requirement of analysis.
4. The questions related to personal information should be are avoided.
5. Only radial question should be asked.
6. Questions with inaccurate response should be avoided.

Wording of questions

Due care should be taken for formulating questions. Response to be based on this fact only. Greater care should be taken while wording of question. In which opinion is to be expressed or actual information is to be provided.

If question is beyond the understanding of respondent than suggestive answer will help than for response.

Guidelines for wording questions

The researcher must pay attention to following point while wording question.

1. Use simple and familiar words to avoid confusion.
2. Formulate question in such a way that will provide required information.
3. The questions having multiple meaning should be avoided.
4. Ambiguity make question complicated so avoid the use of such ambiguous word.
5. Leading question may produce biased answer. Question should be worded in such manner in which untruthful answer may be avoided.
6. Catch words emotional words or danger words should not be used in questions.
7. The use of phrases may reflect up on the understanding of informant. They found to answer the question as they understand and feel about situation. So the phrases should be rarely used.
8. The question must allow all possible response. The provision should also be made for indefinite answers like do not know.:

9. The alternatives should be realistic in multiple - choice question. The make alternative meaningful they are to phrased in terms of concrete situations.
10. The amount of writing required in questionnaire and schedule should be reduced to minimum.
11. A few questions should also be designed which can checked the accuracy and consistency of information.
12. The use of questions which invites response to values should be avoided.
13. Proper justification should be made in case of any unreasonable question.

Construction of Questionnaire

The researcher has to decide composition of questionnaire while taking into consideration the data which is already available on topic and extent of data now required. The process of drafting questionnaire can be divided into following aspects.

(a) Information required.
(b) Type of questionnaire to be used.
(c) First draft
(d) Revising questions
(e) Editing of the questionnaire
(f) Explaining the procedure for its use.

1. The problem formulation is base to develop the questionnaire. The researcher is expected to decide which aspects of the problem are to be dealt with in a particular study.
2. The researcher must also decide wether to use open end or close end question. In case for free response follow-up questions are advised.
3. While drafting questions sequence of the topic should be carefully considered.
4. The questionnaire should be properly scrutinized so as to remove of technical defects.
5. The questionnaire should be pre tested before its launching By Pre-testing of questionnaire unforeseen problem can be stored.
6. The questionnaire should contain simple directions regarding what informant is expected to do.

Sequence of Questions

The sequence of question means the order in which questions are to be asked. Refusal to questions can be avoided by proper arrangement of questions.

Questions should be logically arranged. The opening question should always be of interest to respondent and easy to answer.

The question embarrassing respondent should be normally at the middle or at the end of the questionnaire.

The question requiring personal data of socio economic status should also be put to the end of questionnaire.

The sensitive questions should be asked at the extreme end of the questionnaire. In case of mail respondent interesting questions should be asked in the beginning.

Types of questions

1. **Contingency questions** - A question that is answered only if the respondent gives a particular response to a previous question. This avoids asking questions of people that do not apply to them (for example, asking men if they have ever been pregnant).

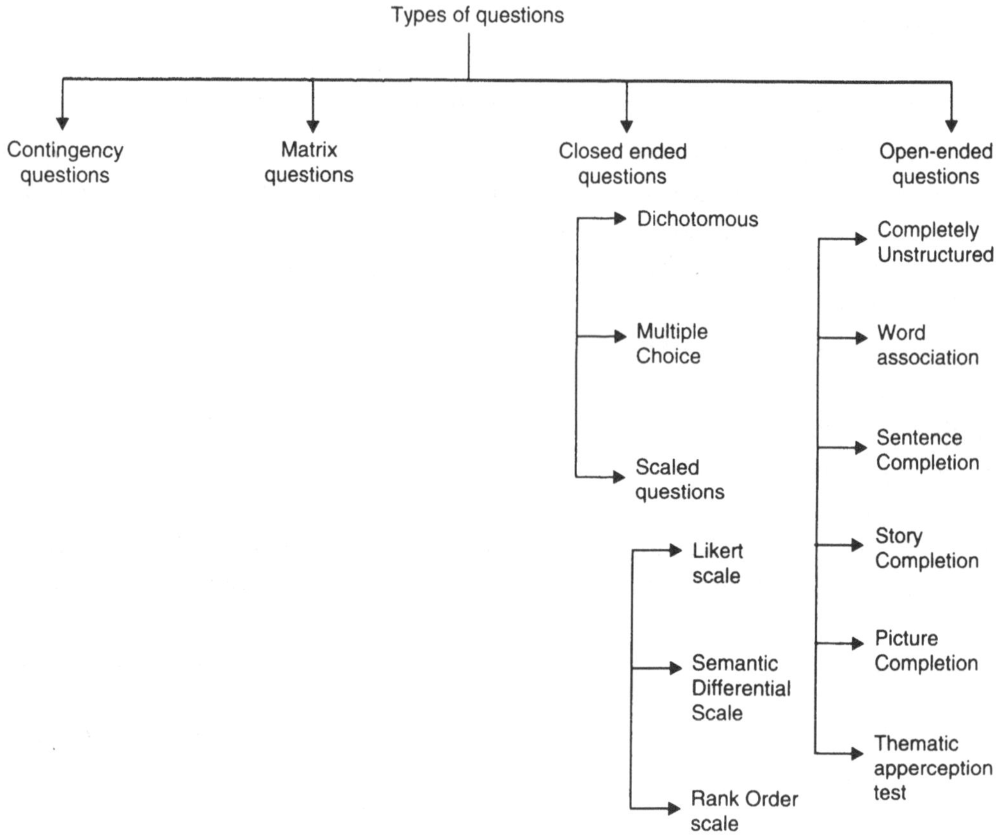

2. **Matrix questions** - Identical response categories are assigned to multiple questions. The questions are placed one under the other, forming a matrix with response categories along the top and a list of questions down the side. This is an efficient use of page space and respondents' time.
3. **Closed ended question**-Respondents' answers are limited to a fixed set of responses. Most scales are closed ended. Other types of closed ended questions include:
   - **Dichotomous questions-**The respondent answers with a "yes" or a "no".
   - **Multiple choice** - The respondent has several option from which to choose.
   - **Scaled questions-** Responses are graded on a continuum (example: rate the appearance of the product on a scale from 1 to 10, with 10 being the most preferred appearance). Examples of types of scales include the Likert scale, semantic differential scale, and rank-order scale.
   - **Likert scale-** A Likert item is simply a statement which the respondent is asked to evaluate according to any kind of subjective or objective criteria; generally

here the level of agreement or disagreement is measured. The format of a typical five-level Likert item is:

1. Strongly disagree
2. Disagree
3. Neither agree nor disagree
4. Agree
5. Strongly agree

Example for Likert scale:

How much do you agree with each of the following statement?

Particulars	Strongly Agree	Agree	Disagree	Strongly disagree
My manager provides constructive criticism	☐	☐	☐	☐
Our medical plan provides adequate coverage	☐	☐	☐	☐
I would prefer to work longer hours on fewer days	☐	☐	☐	☐

- **Semantic differential Scale-**Semantic differential scale is a type of a rating scale designed to measure the connotative meaning of objects, events, and concepts. The connotations are used to derive the attitude towards the given object, event or concept. Example:-

  How would you rate your competency?
  - Good
  - Average
  - Bad
- **Rank order scale-** The respondent is presented with several items simultaneously and asked to rank them (example: Rate the following advertisements from 1 to 10.). This is an ordinal level technique.

4. **Open ended questions -** No options or predefined categories are suggested. The respondent supplies their own answer without being constrained by a fixed set of possible responses. Example of types of open ended questions include:
   - **Completely unstructured -** For example, "What is your opinion of questionnaires?"
   - **Word association -** Word are presented and the respondent mentions the first word that comes to mind.
   - **Sentence completion -** Respondents complete an incomplete sentence. For example, "The most important consideration in my decision to buy a new house is..."
   - **Story completion -** Respondents complete an incomplete story.
   - **Picture completion -** Respondents fill in an empty conversation balloon.
   - **Thematic apperception test-** Respondents explain a picture or make up a story about what they think is happening in the picture.

Formulation of questions and format

If the types of measurements to be made are clearly known, then the researcher can formulate necessary questions relating to each measurement. While formulating a question, its content should be such that each respondent grasps the intention of the question very quickly as originally thought by the researcher. This will minimize the distortion of the research focus. Also, the format of each question in relation to that of the overall questionnaire should be decided.

The format of a question can be of three types : open-ended questions, close-ended questions and structured questions.

Open-ended questions : In this type of question, the interviewer writes the answer of the respondent verbatim. In this type, there is no pre-coded classification of answers to the question.

In a survey to find customer opinion on the interest rate of housing loan, a sample open-ended question without classification is as :

What are the reasons for availing housing loan?

The possible answer for this question may be :

Mainly to avail tax benefit and inculcate saving habit in me.

The open-ended questions may bring unconventional answers which are not known to the researcher. Generally, this type of question is used for probing type question, introductory question to get opinion about the components of the system of study, sensitive question, etc. If the number of alternative answers is many, then one can resort to this type of question. This is alternatively known as unstructured question.

Though it has certain benefits, the answers to unstructured questions may not be realistic as expected if the respondents are not articulate and educated. In telephone interviews, interviewers find it difficult to write verbatim of the answers given by the respondents to open-ended questions.

Close-ended questions

Close-ended questions are structured questions. It has two formats :

1. Questions with multiple responses out of which the respondents have to select one or more choices.
2. Questions with rating scale with discrete responses or continuous range.

A sample format of the first type is as shown below :

What is the number of dependents with you ?

(a) None (b) One
(c) Two (d) Three
(e) Four and above.

A sample format of the second type is demonstrated based on the joint family system as follows :

Question : "Joint family system leads to better GDP of a nation." Please tick one of the following :

(a) Strongly agree (b) Moderately agree
(c) Strongly disagree (d) Moderately disagree
(e) Neither agree nor disagree.

An extreme form of close-ended question has only two responses. The respondent has to select one of these two responses which is demonstrated with reference to a study on consumer survey. This question has dichotomous responses. For example,

Have you used any one of our company's products?

(a) Yes (b) No

A Sample Questionnaire

A study for a tele-services company to find the expectations of customers using telephone booths at Hyderabad and their profiles. The format of the questionnaire used in this study is presented below :

Questionnaire

Study on customer expectations and profiles of pco booths at Hyderabad

Address of Telephone Booth
Customer's personal profile

1. Name :

2. Age :
   (a) Up to 17 yrs (b) 18-24
   (c) 25-40 yrs (d) 41-45 yrs
   (e) 46-55 yrs (f) More than 55 yrs.

3. Genger :
   (a) Male ( ) (b) Female ().

4. Monthly household income
   (a) Less than Rs. 10,000 (b) Rs. 10,000 - 20,000
   (c) Rs. 20,000 - 30,000 (d) Rs. 30,000 - 40,000
   (e) More than Rs. 40,000

5. Occupation
   (a) Service sector (b) Government
   (c) Public (d) Private
   (e) Business (f) Student / housewife
   (g) Others (specifty).
   If business, tick one of the following
   (a) Trading (b) Manufacturing
   (c) Retail (d) Service industry
   (e) Others (specify).
   Are you operating from Hyderabad?
   Yes ( ) No ( )
   If None, mention your base city ....

6. Qualification
   (a) School (b) Graduate
   (c) Postgraduate (d) Doctorate
   (e) Others

7. Ownership details : (Please specify the brand.)
   Two wheeler :
   Four wheeler :
   Cellphone : Roaming ( ) STD ( )
   Credit card :

Date profile :

1. How often do you make a call at a PCO? (Please tick the cell.)

No. of times

Periodicity	1	2	3	>3
(a) Daily				
(b) Weekly				
(c) Monthly				

2. What is the type of call made by you?

   (a) Personal (b) Business (c) Both

3. What are reasons for using PCO?

   ....

4. Do you come to this booth all the time?

   (a) Yes, why? ....

   (b) No, why? ....

5. Do you have a telephone/cellphone?

   (a) If yes, then to to item 6.

   (b) If no, why?

   ....

   Go to item 7.

6. Do you have an STD/ISD connection at home/office ?

   (a) If yes, why do you come to PCO booth to make calls?

   ....

   (b) If no, why?

   ....

   Go to item 8.

7. Are you planning to get a telephone service along with a STD/ISD facility in the near future?

   (a) If yes, why?

   ....

   (b) If no, why?

   ....

8. Are you aware of the charge rates at different timings?
   (a) If yes, mention them ....
   (b) No

9. At what time do you make STD calls?
   (a) Morning (b) Afternoon
   (c) Evening (d) Night

10. At what time do you make ISD calls?
   (a) Morning (b) Afternoon
   (c) Evening (d) Night

11. What is your average weekly spending on PCO calls?
   (a) Less than Rs. 50 (b) Rs. 50-100
   (c) Rs. 100-250 (d) Rs. 250-500
   (e) Rs. 500-1000 (f) More than Rs. 1000

12. Are you satisfied with the billings at PCO ?
   (a) Yes ( ) (b) No ( )
   If No, give reasons : ....

13. What are your general expectations from PCO billings?
   ....

14. Have you encountered any PCO booth with tampered billings?
   (a) Yes ( ) (b) No ( )

15. Rank your order of preference of expectation from 1-6 at PCO booths?
   (a) Good aeration ( ) (b) Privacy ( )
   (c) Good seating facility ( ) (d) Recreation ( )
   (e) Accurate billing ( ) (f) Exact changes ( ).
   Any other, specify : ....

16. "Privacy of conversation at PCO booths can be guaranteed."
   (a) Strongly agree (b) Moderately agree
   (c) Strongly disagree (d) Moderately disagree
   (e) Neither agree nor disagree

17. "Some may feel that coming to PCO booths to make calls is below the dignity of a person."
   (a) Strongly agree (b) Moderately agree
   (c) Strongly disagree (d) Moderately disagree
   (e) Neither agree nor disagree.

18. List the inconveniences faced by you at PCO booths.
   ....
   ....
   ....

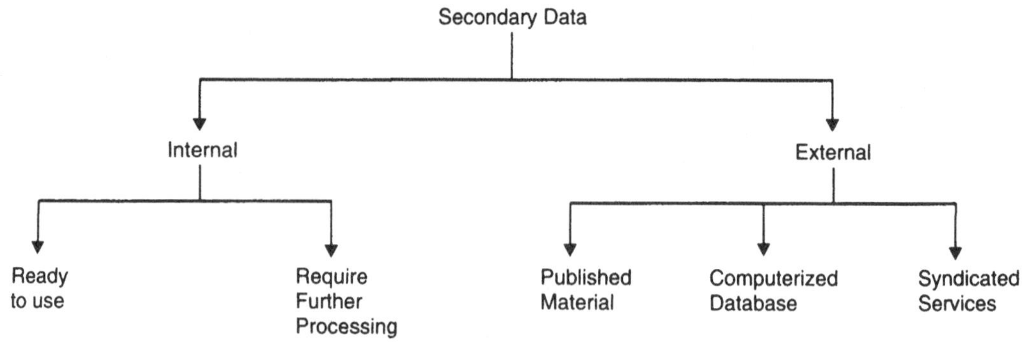

Secondary Data

Secondary data is data which has been collected by individuals or agencies earlier and are compiled in the form of reports, manuals, etc. Thus, secondary data is the data which already exists and is useful for the researcher for his study.

Secondary sources of information can be divided into two categories :

Internal Sources and External Sources

Internal Sources include

–Accounting records

–Marketing records

–Production personnel

–Information system pesonnnel

–Sales force

–Other expert employees

External Sources include

–Govt agencies

–Libraries

–The Internet

–Vendors

– Producers

–Books and Periodicals

–Media Sources

–Trade Associations

–Commercial Sources

(A) Internal Sources :

(i) Ready to use– Those sources which can be used as it is. Like, Sales data, financial data, transport data, storage data, etc.

(ii) Require further processing– Those data which are available but need some more processing before being considered for the study.

(B) External Sources

(i) **Published data-**Those data that are published. The sources of published data are General Business Sources and Govt Sources.

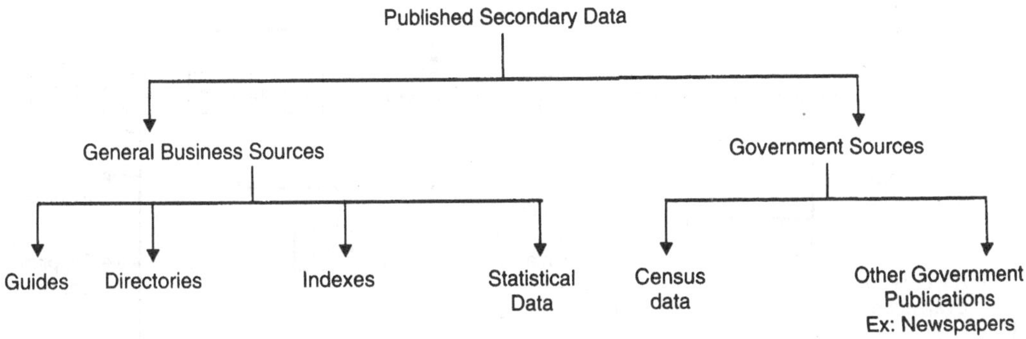

(ii) Computerized databases

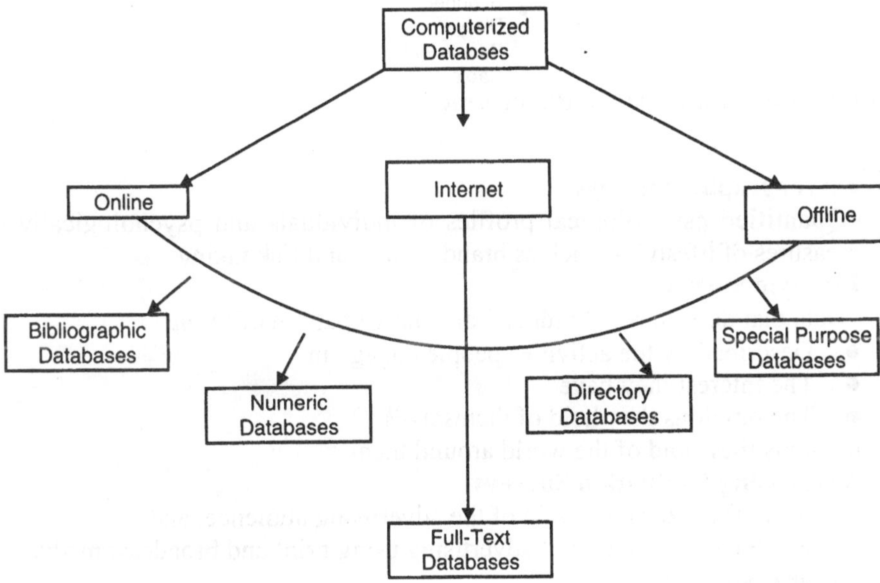

Syndicated Sources (Services)

Firms that collect and sell common pools of data designed to serve information needs shared by a number of clients, including competing firms in the same industry are known as syndicated sources.

Classification of Syndicated Sources

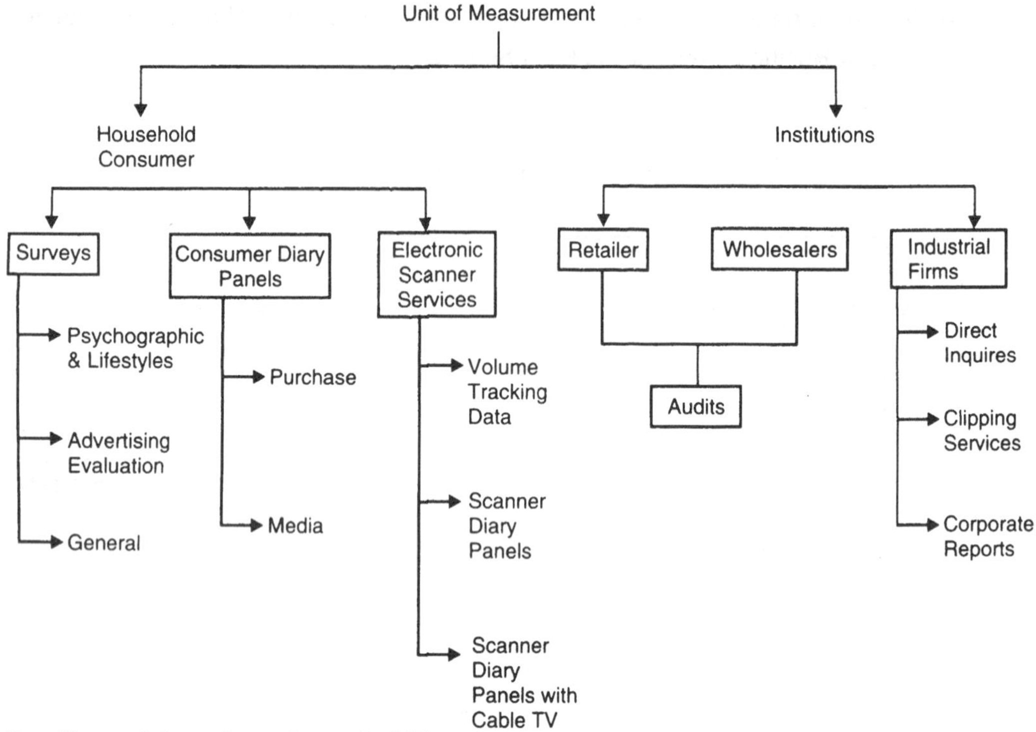

Syndicated data from household/consumers

Surveys

- **Psychographic Surveys**
  –Quantified psychological profiles of individuals and psychologically based measures of lifestyles, such as brand loyalty and risk taking.
- **Lifestyle Surveys**
  –Data that describes individuals' distinctive pattern of living
  - Described by the activities people engage in
  - The interest they have
  - The opinions they hold of themselves

The opinions they hold of the world around them

- **Advertising Evaluation Surveys**
  –Measure the size and profile of the advertising audience, and
  –Assess the effectiveness of advertising using print and broadcast media
- **General Surveys**
  –Conducted for a variety of other purpose (i.e., examination of purchase and consumption behaviour)

Consumer Diary Panels

- Made up of a group of individuals and households that record their purchases and behaviour in a diary over time.
- Frequent attempts are made to maintain the representativeness of the panel in relation to the population (on the basis of demographics).

Diary Puchase Panels

- Data-gathering techniques in which respondents record their purchases in a diary.

Diary Media Panels

- Data-gathering techniques composed of samples of respondents whose television viewing behaviour is automatically recorded by electronic devices.
- Supplemented by the purchase information recorded in diary

Electronic Scanner Services

- **Scanner Data**

  –Data obtained by passing merchandise over a laser scanner that reads the UPC (Universal Product Code) from the packages.

  –Volume-tracking Data

  - Scanner data that provides information on purchase by brand, size, price, and flavour or formulation

  –Scanner Diary Panels

  - Panel members issued an ID card allowing panel members' purchases to the linked to their identity

  –Scanner Diary Panels with Cable TV

  - Combination of scanner diary panels with manipulations of the advertising that is being broadcast by cable television companies.

Syndicated data from Institutions

- **Retailer and Wholesaler Audits**

  –Collect of product movement data for wholesalers and retailers

  –Periodic audits may be

  - A physical count of inventory, or
  - May be managed through a link to the scanning process

  –Track inventory flow, current inventory levels, and the impact of both promotional and pricing programs on inventory levels

- **Industrial Reserch Services**

  –Secondary data derived from industrial sources and intended for Industrial use

QUESTIONS

1. (a) Define and classify secondary data. Discuss the process of evaluating secondary data.

   (b) Discuss observation and interview method.
2. Explain the deference between questionnaires and schedules. Describe the contents in construction of questionnaire.
3. Enumerate the different methods of collecting data. Which one is the most suitable for conducting enquiry regarding family welfare programme in India?
   Explain its merits and demerits.
4. Examine the merits limitations of the observation method in collecting material. Illustrate your answer with suitable examples.
5. "Experimental method of research is not suitable in management field." Discuss, what are the problems in the introduction of this research design in business organisation?
6. Explain the meaning and definition of interview and dicuss the types of interview.

■■■

CHAPTER 5

LEVELS OF MEASUREMENT AND SCALING

After going through this chapter, you will be able to know :

- Measurement Levels/Scales
- Error in Measurement
- Realiability Validety and Practicality
- Scaling Techniques
- Reliability Test
- Paired Comparision
- Likert Scale
- Thurstone Scale
- Continuous rating scale
- Line marking Scale
- Intemised rating scale
- Guttman Scales

A common factor in research is the attempt to have respondent communicating there response and

evaluating it in measurable form. In this respect market research have develop rage of scales. Each of those has unique properties. It is important for analyst to realized that they have widely deleing measurement. properties.

Some scales are at very best limited in their mathematical properties to the extent. that they can only establish an association between variables.

Other scales have more extensive mathematical properties and some hold out possibility of establishing cause and effect relationship between variables.

This measurement process gives scope for providing meaningful information for decision making.

Measurement in the broadest sense, can be defined as *the assignment of numerals to objects or events according to specific 'rules'*. The fact that different 'rules' may be used to assign numerals leads to different kinds of scales. There is thus a need to make explicit three aspects of measurement.

(a) The various rules for the assignment of numerals

(b) The mathematical properties of the resultant scales

(c) Statistical operations applicable to measurements using each type of scale.

Measurement may be defined as the assignment of numbers to the characteristics of objects according to levels. The level of measurement being a function of rules under which the numbers are assigned.

There are four characteristics of the number system,

1. Each number in unique 0-9
2. Ordering of number is given by convention 1, 2, 3.
3. Define with equal dillerence 3-2=1, 5-4=1
4. Define with equal ratio - $\frac{10-5}{2}=\frac{6-3}{2}$

Characterstics of measuring scales

Scale	Basic empirical operations	Permissible descriptive statistics
Nominal	Determination of equality	Mode
Ordinal	Determination of greater or less	Median, percentiles
Interval	Determination of equality of intervals or differences	Mean, standard deviation, variance
Ratio	Determination of equality of ratios	Coefficient of variation

Characteristics of a good scale

a. The ability of the scale to differentiate among products.
b. Reliability of use (reliability as an upper bound on discriminative power)
c. Sufficient alternatives; Room to discriminate between products
d. Avoidance of end and ceiling effects
e. Anchor considerations : Is the frame of reference provided ? Obvious? Are the anchor terms used appropriate?
f. Attribute considerations (semantics) : Is the attribute defined for the subject? Does the attribute cover one or many sensory attributes/concepts?
g. Bias considerations : Should numbers be avoided entirely in favour of lines and/or boxes?

Scale Evaluation

Scale is evaluated through reliability and validity.

Reliability of Scale

The following methods can be used to test the reliability of the scale.

Split-Half Method

In this method the scale may be divided into two equal parts. Then each part is taken as a whole (i.e. as a complete scale) and measurement is made is separately. The correlation between the two scores obtained in this way is estimated and if the degree of correlation is high the scale may be regarded as reliable.

Multiple Form

According to this technique two types of scales are constructed and they are applied to the same population. In case the two results are more or less similar he scales may be regarded as reliable.

Retest Method

In this method the same scale is applied twice to the same population for the same objectives and if the two results are similar, the scale is regarded as reliable.

Validity of Scale

There are at least four different criteria for measuring the validity.

Known Groups

In this method the scale is applied to the known category of people. Then, the result obtained is compared with the known facts. If they are similar the scale is considered to be valid.

Logical Validity

The scale must conform to common sense reasoning and therefore is partially subjective. But still in some very clear cases this crude method may be useful.

Opinion of Jury

This method depends on the considered opinion of many jurists who will have no bias. If several jurists give the same opinion the method of measurement may be considered as to be valid.

Independent Methods

A thing may be separately measured by separate independent criteria and if the results are similar the scale may be said to be valid. However, these methods of testing the validity of scale are not free from flaws. Hence they have to be used very cautiously.

The scaling method is used when the researcher wishes to utilize simultaneously a number of observations on a respondent. Where a scale is developed on the basis of objective criteria the validity can be judged by testing the criteria itself. The validity may be for the future or for the present.

Measurement Levels/Scales or Types of Scales

Levels or scale of measurement is a Yard stick. It is just like an instrument for finding weight or volume. The scale does not possess the properties associated with most physical measures. Scales are the methods most often used to turn responses obtained in surveys into numerical indices for use in analysis. In conclusive research we rely on quantitative techniques to express in numeric terms the difference in responses. Hence, a scale is used to represent the item being measured in the spectrum of possibilities. The values assigned in the measuring process can then be manipulated according to certain mathematical rules.

There are four main types of scales. Distinguishing among these scales is important especially in terms of differential application of statistical techniques.

The scales a can be classified into following categories.

1. Nominal categorical assignment of numbers to items, as if merely labeling (e.g. numbers on a team member's uniform).
2. Ordinal rank ordering items, numbers represent relative ranks of objects being scaled.
3. Interval assignment of numbers on the basis of differences (arbitrary zero point).
4. Ratio assignment of numbers on the basis of ratios (i.e., magnitude; the zero point becomes meaningful).

1. Nominal Scale

This the crudest of measurement scales. It classifies individuals companies, products brands or other entities into categories where on order in implied.

In this scale numbers' are used to identify or categorization and does not place the entity along a continuum.

It evolves a simply count of frequency of the cases assigned to the various categories, and if desired numbers can be nominally assigned to label each category.

The example is given below.

Which of the following items do you tend to buy at best once per month (Please tick)

Soya bin Oil ☐	Dry Fruits ☐	Instant coffee ☐
Basmati Rice ☐	Condense Milk ☐	Liquid Soap ☐

Nominal scale is least powerful level of measurement. It only measures of average which can be used in the mode, because this is simply a set of Frequency counts.

Hypothesis test can be carried out on data collected in the nominal from. A nominal scale simply describes differences between things by assigning then categories. It can tell nothing about the form of relationship, it is not capable of establishing cause and effect relationships.

2. Ordinal Scale

The ordinal Scale indicates the relatives position of two or more objects. Ordinal scale involve the ranking of individuals attitude or items along the continuum of the characterstics being scale. Preference order represent ordinal scales and are frequently used in research relating to qualitative phenomena.

For example- If researcher ask consumer to rank 5 brands of toothpest in order of preference he/she might obtain response like those in table below :

An example of an ordinal scale to determine consumers preferences among 5 brand of toothpest

Order Preference- Brand

1. Colgate
2. Pepsodent
3. Vico
4. Babul
5. Miswak

From such table the investigator knows the order of preference but nothing about how much more one brand is prefered to another: there in no information about the interval between any two brands. Ordinal scales only permit the ranking of items from highest to lowest. It is possible to test order correlation with rank data. In the example given here the investigator might wish to measure similarities and differences in ranking of toothpest brands. The resultant coefficient takes a value in the range of 0 to I. A zero would mean that there was on agreement between two groups and I would indicate total agreement. It is more likely that an answer somewhere between these two extremes would be found.

3. Interval Scale

The interval scale has all characteristics of the ordinal scale and in addition the units of measurement or intervals between successive positions, which are equal. The units are equal only in so far as one accepts the assumptions on which rule is based.

It is only with an interval scaled data that researcher can justify the use of the arithmetic mean as the measure of average. The interval scale has equal units of measurement thus making, it possible to interpret not only the order of scale scores but also the distance between them.

For example investigator scale brands Colgate. Pepsodent and Vico on an interval scale regarding the buyers degree of liking of the brands. Colgates recorves the highest liking score of 6 Pepsodent 3 and Vico receives 2. First the liking of for brand Colgate in move faverable than that of brand pepsodent. Second the degree of liking between colgate and Pepsodent is greater than for liking between Peposent and Vico.

Interval scale provides move powerful measurement than the ordinal scales as it incorporate the concept of equality of interval. The statistical tools range, mean and standard deviation are used in research studies , especially in collecting attitudinal and overall brand rating information.

4. Ratio Scales

The highest level of measurement is a ratio scale. This has the properties of an interval scale together with a fixed origin or zero point. They have order, distance and unique origin.

Examples of variables which are on the ratio scale includes weight lengths and times. Ratio scale permit the researcher to compare both differences in scores and the relative maginitu of scores. For instance the difference between 10 an 20 minutes in the same. As in 20 and 30 minutes and 20 minutes is twice as long as 10 minutes.

Rario scales are normally not used in sociological or management research. as they can not go beyond the interval level of measurement. Researchers in Physical Sciences have the advantage to describe variables in ratio scale form but the socio-managerial research is limited to describe variables in interval scale. All statistical operations can be performed on ratio scales.

Difficulties in scaling

1. In the application of scaling it is almost impossible to assess its validity directly.
2. A scale is applicable only to a particular group or class
3. Quantitative measurement is not applicable to intangible qualitative data.
4. There is no universally recognised measuring technique under scaling.
5. The scores of the scale cannot be correctly tested because of the intangibility of the social phenomena.
6. The social phenomena cannot be experimented in a controlled way. Hence scale cannot measure all the causative variables involved.
7. There is often a dichotomy between the express attitude and overt action. Hence even a valid scale may not remain valid during the period of analysis.

Error in Measurement

The research study should be free of any error in measurement. The researcher should be aware of causes of such errors in measurement, explained as under-

1. Respondent

The respondent may not express negative feelings or may not accept his poor knowledge. This may result in expressing opinion without sufficient ground. Factors like stress may limit ability of respondent to respond accurately.

2. Situation

This factor is main hurdle in correct measurement . If interviewer is exposed to embrancing situation it will have effects on his response. If he is interviewed in presence of person where he feels lack of anonymity than he may be reluctant to express freely.

3. Interviewer

The interviewer can discourage the response, committed by measurer by incorrect coding, wrong tabulation or by mistake in calculations.

4. Measuring Instrument

Defective measuring instrument will lead to errors. Poor sampling of universe. use of complex words inadequate space to replies all these make measuring instrument defective which results in measuring error.

Types of Errors

Stage of Research	Type of Error
1. Concept and hypotheses hypothesis construction (including choice of operational definitions)	Lack of face validity.
2. Construction of research instrument (questionnaire)	Lack of reliability (faculty or ambiguous working on questions.
3. Sampling	Lack of external validity (sampling error).
4. Data gathering	Error due to failure to control : (a) Environment; (b) Personal characteristics of respondent (fatigue, etc.); (c) Relationship between respondent and researcher; (d) Defects of research instrument (faculty sound recording, equipment failure, etc.); (e) Interviewer misunderstood answer.
5. Coding	Incorrect information recorded due to missing data, illegible data, or simple coding error.
6. Data analysis	Misuse of statistics or faulty interpretation of data.

Test/Retest/Reliability

Since reliability by definition means consistency in the scores of a single measure, rather than identical scores on two alternate measures, it would seem better to test for reliability through repeated applications of the same measure. This is a view shared by

Siegel and Hodge, who say, "In order to make any substantial progress on questions of reliability one needs, of course, repeated measurements."

In order to understand fully all of the problems associated with repeated application of the same measurement instrument in social research, we need to examine in more detail exactly how a measuring instrument can be unreliable. If you use your bathroom scale repeatedly over a period of two months, weighing every day, and find that at the end of two months your recorded weight is three kg heavier than it was two months previously, this is not necessarily evidence of lack of reliability. You may have gained three kg over the course of two months, so that the accurate scale must show you to be heavier. A reliable scale will show a change in the value of the characteristic being measured whenever such a change occurs. It just should not show a change when none occurs.

Fresh students of research methods sometime think that in order to be reliable a measuring instrument must show consistent value over each measurement period, but this is true only if no change actually occurs. We have all seen a broken thermometer that shows the same temperature in January as in August; though it shows consistent scores we certainly do not consider it reliable. The problem in social research lies in determining, whenever repeated applications yield inconsistent scores, whether the value of the concept really changed, or whether the measuring instrument is unreliable.

Reliability Validity and Practicality

The reliability and validly are the important for testing of sound measurement.

A. Reliability Test

Reliability means stability and consistency. It is test of the sound measurement. Reliability provides same results in the repeated application to the same event.

The stability which in the main aspect of reliabity is concerned with getting consisting results with repeated result with same instrument.

The equivalence is the second aspect consider the fact, as to how error may enter by different investigators conducting study. The best way to compare the observations on same events.

There are three methods why which reliability can he tested

(i) Test-retest
(ii) Spilt- half
(iii) Alternative form

(i) Test- retest : in this method same measure is applied twice to same individuals at two different points and the resulting scores to be correlated. The higher co-efficient of correlation is the higher reliability and measure.

(ii) Split-half : under- this method large set of questions are divided in to sub-measures. The technique of spliting is used odd and even numbers for creating sub-measures. These both sub measure so created are administered to respondents. The scores of sub measures are then correlated and this co-efficient is called co-efficient of internal consistency to be taken as measure of reliability.

(iii) Alternative method : this method consist to forms of measures to be administered to same individual. The result of the two forms are correlated to decide the degree of reliability

This method is used to decide the impact of certain change programme, like change in attitudes or in values.

B. Validity Test

Validity is second test of sound measurement. It refers to the exigent installment. It also refers to measurement process free from systematic and random errors outcome of systematic error therefore for the validity of measure systematic and random bias are to be zero. If measures are not reliable than it can not be valid.

The validity can be measured by three types (i) content validity (ii) construct validity (iii) criterion validity.

(i) Content Validity - It determine primary judgment. It can also be decided on the bases of using expert panel who will judge how measuring units meets the standard but have the numerical expression, which is not possible to be measured by developing questionnaire of 10 questions. So the validity will be decided using expert panel.

(ii) Construct validity - It relates to understanding to measurement instrument providing adequate coverage of topic under study to construct validity. Which is the degree to which scores of test can he accurate for, by the explanatory construct of sound theory. It could be assessed by developing measures by two constructs and ascertaining the relationship among the set of measures.

(iii) Criterion validity - It relates to ability to predict or estimate outcome of current condition. It reflects the success of measures used for estimating purpose.

The criterion validity must have qualities like relevance freedom, from bias. reliability and availability.

The criterion validity broadly relates to productive and criterion validities. They referred to usefulness. of test predicting and of closely related other measures.

Scaling Techniques

In all cases the measurement instrument angle to be practical. In this test viability can be judge intermes of convenience economy. Interpretation economic validity is required in ideal research project, that which unit can of afford. Convenience test suggest that measuring instruments which are going to be adopted is the most important factor. Any body other than designer should able to interpret the results.

Reliability Test

The various types of scales used in research which can be broadly classified as comperative and non-comperative scales.

In comparative scale respondent is asked to compare one brand or product or factor to other one. On non-comparative scaling respondents needs to evaluate a single brand or product or factor. Their evaluation is independent of the product/brand/factor which research is studying. Non-comparative scaling this is the move widly used. Type of scale in commercial marketing research studies is comparative scales.

1. Paired Comparision

In market research sometimes researchers wish to find out which are the most important factors in deciding the duration for product. It means researcher wish to know factors responsible for preventing the widespread adoption of the product.

In this method respondent can express his attitude by making choice between two objectives. The attitude is structured in three integrated components of belief feelings and behavioral intentions. Attitude can be measured by asking question to respondent as to how he feels about object.

There are three steps involved in attitude measurement. First one should develop a set of statements related to subject enquiry. Second to choose the statement to be included in final scale. Third the scale once formed should be tested for its reliability and validity. If scale satisfy these criteria than it can be used in the survey.

For example - farmer response to newly design farms equipments by Agro. tech co. It was observed that the fallowing factors were responsible for negative attitude of farmers towards new farm equipment

1. Expensive
2. Risky new technology
3. Transport Difficulties
4. Not useful for of inter-cropping
5. Non-availibity of financial assistance

The organisation wants to know which factor is for most in the mind of farmer which prevents widespread adoption , so that equipment can be redesign.

The process of ranking objection from most to least important is best approached through the questioning technique know as paired comparison. Each of the objections is paired by researcher so that with 5 factors as in this example there are 10 pairs which is given in the formula.

$$N=\frac{(n-1)}{2}$$

Where N = number of judgement

n = number of objectes to be judge

$$N=\frac{5(5.1)}{2}=\frac{5\times4}{2}=\frac{20}{2}=10$$

N = 10 pairs

In 'paired comparisons' every factor has to be paired with every other factor in turu. However only one pair in even put to the farmer at one time pairs are made as under.

1-2	2-3	3-4	4-5
1-3	2-4	3-5	
1-4	2-5		
1-5			
4 +	3 +	2 +	1 = 10

Thus in paired comparison data may be treated in several ways. If brand A is preferred to B and B to C than A will be consistently preferred to C

Suppose the factors explained earlier for farmers response labels as below for easy understanding.

A1 - Expensive

B2 - Risky new technology

C3 - Transport difficulties

D4 - Not useful for intercropping

E5 - Non availability of financial assistance.

This data is arranged in to following table

assume 100 farmers were interviewed for their response which is arranged to in the grid below. The figures are arranged top to side, for example -that 42 out of 100 said equipment is expensive was greater deferent of the fact that , it was not useful for inter-cropping ,was more important than of non-availability of financial assistance.

	A	B	C	D	E
A	42	60	90	87	50
B	50	80	83	88	18
C	20	50	62	84	40
D	24	16	51	50	13
F	1 7	38	50	49	10

If the above table is carefully read it can score that rank order of the factor is most Important

A - Expensive

B - Risky new technology

C - Transport

D - Inter cropping

less important E - Non availability of finance.

It should be clear from the results drawn from the above table that comparison scale gives ordinal data.

Paired comparison technique

If there are n stimuli, then there will be $2^{n}C_2$ pairs of 2 stimuli. For each pair of stimuli (i, j), if a respondent feels that the stimulus i (row i) wins over the stimulus j (column j) (the stimulus i dominates the stimulus j), then mark the respective matrix element with 1. The diagonal elements of the matrix have no values. A sample set of original data for five brands is shown in Table 1. Later, the matrix can be permuted such that the upper diagonal entries are with 1. Then the corresponding order of the row numbers indicates stimulus rank order. These are illustrated in Tables 2 and 3. From Table 3, it is clear that the stimulus rank order is 2, 1, 4, 3 and 5. The total number of 'one entry' of the rows in Table 3 decreases from top to bottom. The data in Table 3 are characterized as hypothetical data because the respondent is considered to be transitive. The meaning of the respondent being transitive is that there is no cycle in the data. Sometime, in the matrix, a cycle may be formed, shown as follows, if the entries for the cells, say (2, 5), (5, 3) and (3, 2) are assumed to be 1.

Brand 2 → Brand 5 → Brand 3 → Brand 2

Existence of such cycle signifies the fact that the respondent is non-transitive. Under such situation, some modifications to the original data should be carried out to make the matrix to the format, as shown in Table 3, in which all the upper diagonal cells are with 1 and all the lower diagonal cells are with 0.

In addition to the transitive property of respondent, one should check the consistency of marking of the stimulus *i* (row *i*) wins over the stimulus *j* (column *j*). This can be taken care by replicating the data using several respondents. Such replicated data can be used to arrive the proportions of preference for different pairs of stimulus *i* over stimulus *j*.

Table - 1

Paired Comparison Responses from a Respondent

Brand		1	2	3	4	5
	1	-	0	1	1	1
	2	1	-	1	1	1
Brand	3	0	0	-	0	1
	4	0	0	1	-	1
	5	0	0	0	0	-

Table - 2

Row Permutation of Table 1

Brand		1	2	3	4	5
	2	1	-	1	1	1
	1	-	0	1	1	1
Brand	4	0	0	1	-	1
	3	0	0	-	0	1
	5	0	0	0	0	-

Table - 3

Column Permutation of Table 2

Brand		2	1	4	3	5
	2	-	1	1	1	1
	1	0	-	1	1	1
Brand	4	0	0	-	1	1
	3	0	0	0	-	1
	5	0	0	0	0	-

Let us assume that this situation is replicated two more times (second and third replications) in addition to the data in Table 1 (first replication) and the corresponding permuted data be, as shown in Tables. 4 and 5.

By adding the data of the corresponding cells of Tables 3 - 5, the total frequency of preferences for the upper-diagonal cells based on the three sets of replicated data are, as shown in Table 6. The respective proportion of preferences are shown in Table 7. Each cell entry in Table 7 is obtained by dividing the respective frequency in Table 6 by 3 (no. of replications).

Table - 4

Permuted Data of Second Replication

Brand		3	1	4	5	2
	3	-	1	1	1	1
	1	0	-	1	1	1
Brand	4	0	0	-	1	1
	5	0	0	0	-	1
	2	0	0	0	0	-

Table - 5

Permuted Data of Third Replication

Brand		2	1	5	4	3
	2	-	1	1	1	1
	1	0	-	1	1	1
Brand	5	0	0	-	1	1
	4	0	0	0	-	1
	3	0	0	0	0	-

Table 6 Total Frequencies of Row Over Column Preferences for the Upper-diagonal Cells, Using Three Permuted Replicated Data Matrices (Tables 3-5)

Brand		1	2	3	4	5
	1	-	1	2	3	3
	2		-	2	2	2
Brand	3			-	1	2
	4				-	2
	5					-

Table 7 Proportion of Row Over Column Preferences for the Upper-diagonal Cells

Brand		1	2	3	4	5
	1	-	.33	.67	1	1
	2		-	.67	.67	.67
Brand	3			-	.33	.67
	4				-	.67
	5					-

In Table 7, the proportion of preference of the brand-1 over the brand-2 is 0.33. Therefore, the proportion of preference of the brand-2 over the brand-1 is 0.67 (i.e. 1-0.33). Based on this logic, the entire pairwise preference proportions are summarized in Table 8.

Table 8 Proportions of Row Over Column Preferences for all Combinations of Rows and Columns

		Brand				
		1	2	3	4	5
Brand	1	-	0.33	0.67	1.00	1.00
	2	0.67	-	0.67	0.67	0.67
	3	0.33	0.33	-	0.33	0.67
	4	0.00	0.33	0.67	-	0.67
	5	0.00	0.33	0.33	0.33	-

The steps of paired comparison technique are summarized as follows :

Step - 1 : Decide on the number of respondents (subjects) which represents the number of replications of the data matrix. Let it be n.

Step-2 : Obtain the responses for each of the respondents in a separate matrix in the format, as shown in Table 1.

Step-3 : Obtain a final matrix after row and column permutations for each respondent.

Step-4 : Obtain the matrix containing cumulative frequencies of preference for the upper-diagonal cells by adding the corresponding cell entries of all the final matrices obtained in step-3.

Step-5 : Divide the cumulative frequency in each upper-diagonal cell of the matrix obtained in step-4 by the number of replications (n) and treat it as the proportion of preference of the row over the column of that cell.

Step-6 : Find the proportions of preference for the lower diagonal cells. The proportion of preference of row i over column j (i>j), P_{ij} is obtained using the following formula.

$$P_{ij} = 1 - P_{ij}, \quad \text{where } i > j$$

Step-7 : Treat the resultant matrix (Table 8) as the input to the related scale.

2. Likert Scale

A likert scale is what is termed as summated instrument scale. This means that the items making up a liker scale are summoned to produce a total score.

In fact a liken scale is composite of itemized scale. In this each scale item will have five categories with 0 as natural response. This response can be categories in degrees of agreement at five levels like strongly Agree. Disagree, strongly Disagree. Agree or Neither or Don't know.

These responses are equivalent to assigning scores of I to 5 and index can be created by summing scores the scores can be reverse in case of negativity worded statement.

Procedure for Likert Scale

1. Relevant statements being measured are collected and classified either favorable on unfavorable, and nutural items to be Omitted.
2. Pre-test for the respondent expressing opinion of each of the item as per the categories of response in conducted.
3. Response recorded is given numerical Score raging from +2. +1. 0. -1. -2
4. By adding item score total respondent score is obtained.
5. By analyzing pre-test, items to be included final test are decided.
6. The final test is administered to the group of respondent of interest and is scored in the same way as in the pre-test

Table : The Likert Scale

Particulars	Strongly Agree	Agree	Neither Dont know	Disagree	Strongly Disagree
If Government imposes heavy taxes the firms would incrcase te prices of product	1	2	3	4	5
If raw material prices a tell the would reduce the product prices	1	2	3	4	5
The pharma industry spend excessively on advertising	1	2	3	4	5
If Government gives taxation relief to industty, consumer will be benifited	1	2	3	4	5

This scale is commonly used in marketing research. It is treated as yielding data by majority of marketing research. This scale is widely used for its advantages like simpler to construct, provides for expression, for intensity of feeling and as it appears to be more reliable.

3. Semantic Scales

In this scale extensive use of words is made rather than using numbers. A semantic scale. Offer bipolar positions to a single stimulus.

Respondent describe their feelings about the products or brands on the scales with Semantic labels. When bipolar adjectives are used at the end points of the scales. these are termed as semantic differential Scales. It consists and set of descriptive polar opposite rating scales each assumed to be an equal interval ordinal scale.

The scale can have any number of positions, although usually it is an odd-numbered position scale generally seven points.

The semantic scale and semantic differential scale are shown in the following table

Semantic and Semantic differential Scales

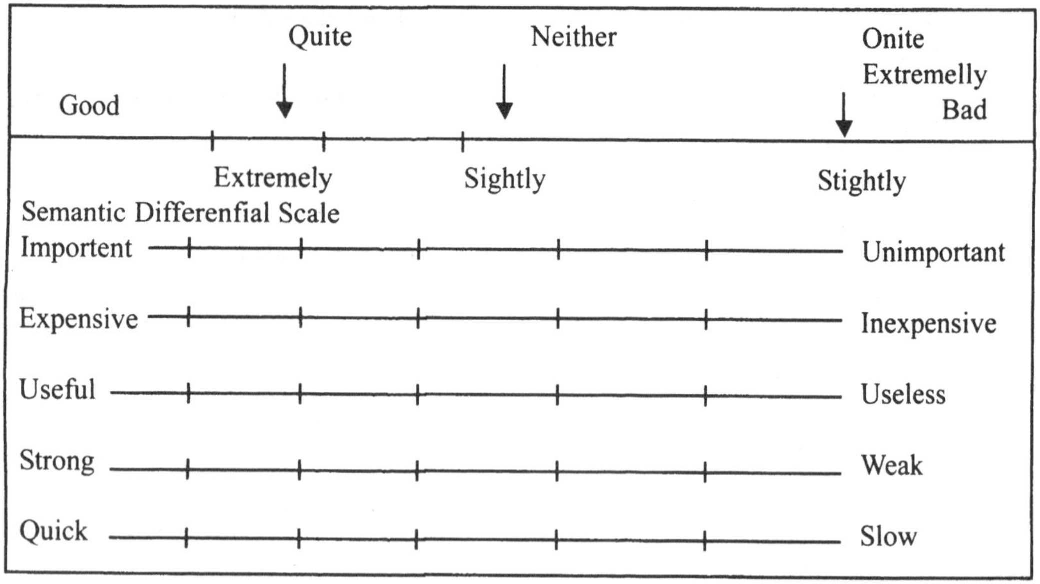

To measure reaction of the attitude towards the product would be calculate by summing up individuals, response as shown the table above. In semantic scale attitude is recorded in 7 semantic labels as Good,(G- extremely, quite, slightly, neither B- slightly quite, extremely) and bad.

In semantic differential scale response is recorded in two opposite rating.

4. Thurstone Scale

The Thurstone scale has been widely used for developing differential scales which is utilized to measure attitudes varied issues like war, religion etc. The following are the factors for consideration for construction of this Scale.

1. Large number of statements are collected relating to enquiry and they should range from one extreme of unfavorable attitude to another extreme of favorable attitude.
2. Thurstone recommended five criteria for selecting statement. That the statement should be brief and rated to attitude of the respondent. This statement's of rejection on acceptance should treated as indicator or respondents response and ambiguity in statement should be avoided.

Based on above productive statements are selected. The scale values are equally spaced. such scales are considered most important when used in measuring single attitude.
The statements in the scale are included in random order and respondents are required to express on all the statements to which he agrees.

The method is not completely objective it involves subjective decision process.

This method is also criticized on the ground that other scales have more information about respondent's attitude.

5. Continuous rating scale

In continuous rating scale respondent is asked to provide for rating by allotting a mark at the appropriate position on a continuous line. The respondent can be shown the card on which scale is recorded during the interviews. Two types of continuous rating scale are show in the following Table

Table- continuous rating scale

Having used the 'k' brand basmati rice for 2 months now how do you to feel bout this brand.

Scale type A

Unfavorable ——— favorable

0 10 20 30 40 50 60 70 80 90 100

Scale type B

Unfavorable ——— favorable

In type A scale response of product from unfavorable to favorable a marked by using numbers on continuous line from 0-100 But in case of type B scale. the respondent score is decided on the basis of dividing the line into various categories as desired. With this respondent will be assigned on the basis of score to the category he/ she belongs.

6. Line marking Scale

In order to measure perceived similarity differences between products, line marking scale is used. It is a form of semantic differential scale as each end of scale is labeled with a word which is opposite to each other.

The respondent in most of the cases find this scale to be a better one. In this scale they are not required to use number like (5,4,3,2,1) where they reflect their attitudes in words. This is a continuous scale.

7. Intemised rating scale

In this scale, number and description, both factors are used. The respondent is provided with a scale having number and corresponding brief description associated with each category. They are asked to select one of the limited number of categories, as indicated in order in terms of the scale position by which product alternative can be studied. In following table the examples A itemised rating scale are give.

Table Itemised rating scale

How easy or difficult did you find 'V' brand wheat flour to cook.

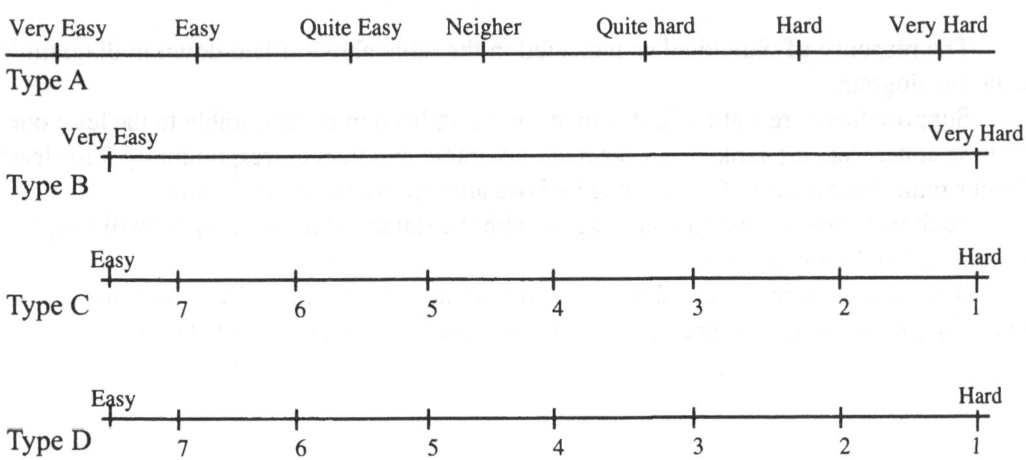

Itemised rating scale can take number of forms as show in above table.

8. Guttman Scales

In case of total attitude is scaled Guttman scale is constructed for this requirement. This scale includes sample of items representing the universe. These items are administered to a sample of persons. The respondent are subject to test the salability of itmes. This type of analysis is know as scalogram. It means question should belong to same attitude dimension.

Example- In order to test ability for Arthmetic the following problems are arranged in order of difficulty

1 5 + 6 =

2. 31 + 39 =

3. 49 + 61-19 =

4. (35 + 19) - 17 =

5. (48 x17) - (28-21+5) =

Any person who can give correct answer to problem 5, can have all problems correct. But one having, 4th problem correct can have problem 3 to I as correct

This can apply to six pallon of answer indecating '+' as correct answer and '-' as wrong answer.

Problems

(1)	(2)	(3)	(4)	(5)	Score
+	+	+	+	+	5
+	+	+	+	-	4
+	+	+	-	-	3
+	+	-	-	-	2
+	-	-	-	-	1
					0

The response of individual as indicated in the table above if laid down in diagram is called scalogram.

Suppose there are 5 attitude statements ranging from most favourable to the least one. If these statements are ranked as I for most favorable and than correspondingly 5 for least favour rank. Take a case of 5 responded whose attitude we want to measure.

Each respondent is asked if he is agree with the statement or not. If agree will be given weight of I and disagree as 0.

If attitude statement fall along unidimensional continuum and the response of the individual to the statement are decided on basis and position in the table below.

Respondents Score and Rank

Respondent						Score of Respondent	Rank of Respondent
	1	2	3	4	5		
A	1	1	1	1	1	5	1
B	0	1	1	1	1	4	2
C	0	0	1	1	1	3	3
D	0	0	0	1	1	2	4
E	0	0	0	1	1	2	4
F	0	0	0	0	0	0	6

From the above table researcher can very well know, the fact that, set of attitude statement falls along an undimensional continuum from least to the most favorable one.

QUESTIONS

1. Explain the scales of measurement. Point out the possible sources of error in measurement. Describe the test of sound measurement.
2. "Scaling describes the procedures by which numbers are assigned to various degrees of opinion, attitude and other concepts." Discuss. Also point out the bases for scale classification.
3. What is the meaning of measurement in research? What difference does it make whether we measure in terms of a norminal, ordinal, interval or ratio scale?
4. Point out the possible sources of error in measurement. Describe the tests of sound measurement.

■■■

CHAPTER 6

PROCESSING OF DATA

After going through this chapter, you will be able to know :

- Presenting of Data
  - Editing
  - Coding
  - Classfication
  - Tabulation
- Components of Table
- Analysis of Data
- Statistical Analysis of Data
- Measures of Central Tendency
  - Mean
  - Median
  - Mode
  - Harmonic Mean
  - Geometric Mean
  - Index Number
- Analysis of Time Series
- Meassure of Dispersion

The data which is collected for the purpose of the study itself cannot reveal any thing. This being a raw data it is required to process and analysed in order to have desired result. The data which is collected can not be directly use for making analysis. Before analysis data is requires to be process.

Data processing is a crucial stage in research. After collecting the data from the field, the researcher has to process and analyse them in order to arrive at certain conclusions which may confirm or invalidate the hypothesis which he had formulated towards the beginning of his research worth. The mass of data collected during the fieldwork is to be processed with a view to reducing them to manageable proportions. Only by such a careful and systematic processing, the data will lend itself for statistical treatment and meaningful interpretation and conclusion.

The processing of data includes editing, coding classification and tabulation. The data which is collected is analyse with an object checking of data, and reducing this data into manageable proportions. Data collected should he organised in such way so that table charts can be prepared for presentation. The processing of data includes various operations which will be explained here.

Once the data begin to flow, a reseacher's attention turnnnnns to data analysis. Data preparation includes editing, coding, and data entry and is the activity that ensures the accuracy of the data and their conversion from raw form to reduced and classified forms that are more appropriate for analysis. Preparing a descriptive statistical summary is another preliminary step leading to an understanding of the collected data. It is during this step that data entry errors may be revealed and corrected. Exhibit reflects the steps in this phase of the resarch process.

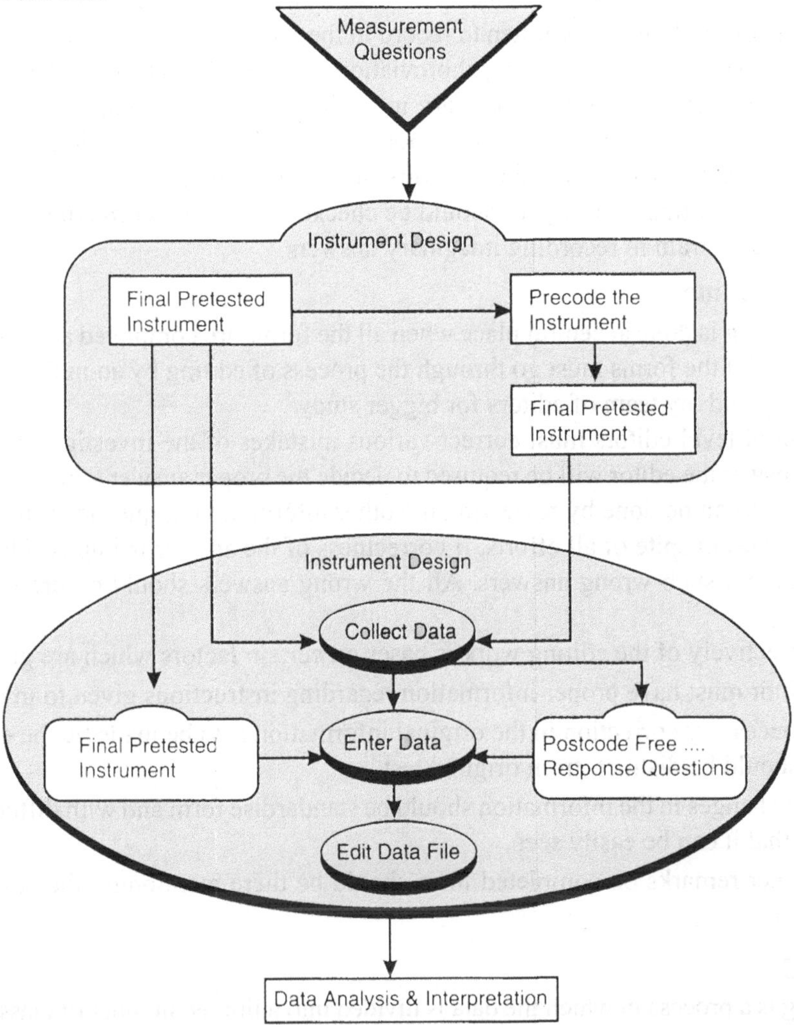

1. Editing

The process of examining the collected data is called editing. This include inspection of data its correction and modification. Therefore editing means scrutiny of the questionnaire and schedule. There is possibility of error and omission in the on the part of investigator so the editing help us to minimise such incorrectness. Editing is done at the time of collecting data.

In case of small study single person can undertake the job of editing where as in case of bigger studies this job can be divided among the various persons.

The scrutiny of the questionnaire or schedule is the main aspect in editing. If the examination of questionnaire and schedule is necessary find out its completeness, uniformity and accuracy.

The process of editing can be conducted at two stages one at field level and another at central level.

A. Field Editing

In field editing completeness of the forms should be checked by persons. It may possible that investigator might have forgotten to record in the information. If investigator recorded information in incomplete form using abbreviations than it should be completed.

This typo of examination is necessary immediately after information is collected by investigator, because the investigator may not be available for clarification latter on. The editor must check the accuracy and uniformity of the answers.

In the field editing investigator should be checked in making corrections at his level. He should also restrain in recording imaginary answers.

B. Central Editing

This editing is done at central place when all the forms are completed and return to the centraloffice. All the forms must go through the process of editing by an individual in case of small study and by team of editors for bigger study.

At central level editors must correct various mistakes of the investigator. In case of gap in the answer the editor will be required to decide the proper answer to meet out the gap in answer. This can be done by reviewing the other information in questionnaire.

Some times in spite of all efforts, if correctness of the answer is impossible than it is safe to strike out such wrong answers. All the wrong answers should be dropped by the editors.

The effectively of the editing work is bases on certain factors which are given below

a. Editor must have proper information regarding instructions given to investigator.
b. If necessary correction in the original information is to be made by the editor than it should be distinct, from original test.
c. The changes in the information should be standardise form and with different color so that it can be easily seen.
d. Proper remarks on completed form should be there mentioning the editing work done.

2. Coding

Coding is a process in which the data is divided into a limited number of classes. In this process all the answers are assigned in any symbol or numeral order. This is done with the help of set rules.

The classes or categories should be reasonable and should be appropriate to the research problem, under study.

The coding must be exhaustive, it means there should be class for each item of the data. For each answer it should be assigned with separate number.

The coding should be based on the fact of mutual exclusivity it means specific answer can be place only in one category.

The coding must observed the rule of single dimension, it means every class in the category set is defined in terms of only one concept.

Coding provides base for analysis. It can be simplified if use of precoded questionnaire is made. The decision of coding should be taken well in advance at the stage of designing of questionnaire. By the investigators time is saved. Standard method should be used in case of hand coding.

The process of coding is hard task. This can be simplified if coding preparation is made prior to designing of schedule or questionnaire. What ever method is adopted for coding the main important fact is that coding errors should be reduced.

3. Classification

Classification is a process in which large data is reduced into homogeneous group if we want to have meaningful relationship. It is a process of arranging data into group based on common characteristics; classification can be done either according to attributes or according to class intervals.

(A) Classification as per attributes

The data which is required to classified as per attributes. It can be either descriptive or numerical. If the data is collected in qualitative means in descriptive form that it can not be measure in quantitative form.

Data collected based on attributes is classified according to attributes only. In attribute classification only one attribute is considered to divide the universe in two classes, one having attribute and other not having the attribute.

In classification based on more number of attributes, than data is divided in number of classes, for example- if attributes AB are studied than their presence is recorded as AB respectively and their absence is recorded in ab respectively. Then we have following classes - AB, Ab, aB, ab.

(B) Classification as per class intervals

The numerical data collected refers to quantitative form, it can be measured through some statistical unit, data related to production, income, age come under this category. This type of data is classified on the basis of classes interval. For example person having income of Rs. 7.000 or Rs. 11.000 per month, group one for 7000 group within Rs. 5.000 - 10,000 can be formed for another for Rs. 11,000 can be formed within 10,000 - 15,000, with formation of these group as it can be seen that every group will have lower and upper limit. This is known as class limits.

The main problem in this classification is that how to make classes, decide limits and frequency in each class.

The processing of data according to class interval requires to decide number of classes normally multiple of 2.5 or 10 are preferred for deciding the number of classes. Some data may have high or low values, so in cases like this open end interval exercise is to be administered.

Now how to decide the limit of the class is also important aspect. In this case midpoint criterion should be considered in which sum of upper and lower limit is divided by 2. The class limit should be located in multiple of 2.5.10, following form.

100 - 500
500 - 1000
1000 - 1500
1500 - 2000
2000 - 2500

The same class limit shown above.

Class should be read as

1 From 100 upto 500
2. From 500 upto 1000
3. From 1000 upto 1500
4. From 1500 upto 2000

In the example given above. Upper limit say in `3' - 1500 as lower limit of 4 now it having value 1500 where it should be included. So under exclusive type of class interval the upper limit as excluded and item with the value less than upper limit are up in the given class interval. So 1500 will be put in class 4 - 1500 to 2000 and not in class - 3, 1000 - 1500 as stated above.

Sometimes the class interval is done with inclusive number.

Class

1 101 - 500
2. 501 - 1000
3. 1001 - 1500
4. 1501 - 2000
5. 2001 - 2500

In this type of class interval the upper and lower limit clearly mention now 1500 will be included in class 3 only.

The third aspect of the class interval is to decide frequency of each class. It is decided on the basis of tally sheet prepared on some mechanical device. In the preparation of tally sheet, it is recorded in multiple of five and this is recorded with four vertical lines with fifth horizontal line. This frequency counting in the tally sheet is show in this table.

Table

Tally Sheet deciding the number of family in income group

Income		Tally	Mark				Class frequency
bellow	-	1,00,000	~~IIII~~	~~IIII~~	~~IIII~~	II	17
1,00.001	-	5,00,000	~~IIII~~	~~IIII~~	~~IIII~~	IIII	24
5.00.001	-	10.00.000	~~IIII~~	~~IIII~~	I		11
10,00,001	-	15,00,000	~~IIII~~	~~IIII~~	III		13
15.00,001 and above	-		~~IIII~~	~~IIII~~		10	
Total							75

4. Tabulation

After collection of row data it is necessary to organised this data in logical order. Arranging this data in from of tables, is known as tabulation. It is process in which summarised data ,is presented in compact form of statistical table. So the presenting row data in table format is the process of tabulation.

The process of tabulation provides base of computation facilitates comparison and detecting errors. By tabulation descriptive statement can be minimised. This process can be conducted manually or mechanically. The availability funds, time and manpower in relation to cost composition are main consideration for method of tabulation.

In case of small investigation , normal tabulation is adopted but for bigger studies mechanical tabulation is advised. Manual tabulation is done by tally sheet or count method.

When bigger studies are conducted, mechanical or electronic devices are used for tabulation. Now the use of computers is very common. In this process the data have to be converted into numeric form for its adoptability on punch cards. Respondents will be given cards and information will be entered in cards so as to do further processing of information.

As explained earlier manual tabulation can be done by making use of direct tally list and tally card sort and count method strip method.

For small group of respondent, having small number of tables, direct tabulation is done. In listing method code response are recorded on large - coding sheet in which code number and question is entered on top of the column.

In card sorting method date are recorded on special card having series of holes. Each hole stand for particular code. These cards are separated and counted and like this code frequency can be found by, act of technique repetitions.

In case of large number of items in big study, project strip method is used. Long strep used in this method , where 160 columns and 320 items can be recorded in one strip.

Classification of Tabulation -

Tabulation can be classified as

A - Simple Tabulation

B - Complex Tabulation

A. Simple Tabulation

It gives information about one or more groups of independent questions. This results, in one way table, provides information of one characteristics of data.

B. Complex Tabulation

In this type of tabulation, the data is divided in two or more categories which gives information regarding more sets of inter-related question. It results in two way or three way tables which gives information about several inter related characteristics of data. This complex table is describe as cross tabulation-

Components of Table

There are set rules for tabulation but due care should be taken on following aspect for constructing table.

1. Each table should have clear number for the purpose of the reference.
2. Every table should also have suitable title this title should be self explanatory.
3. There should be proper heoding to each column and raw of the table in brief.
4. The body of the table contains the numerical information. Data presented in the body is arranged as per the description.
5. The unit of measurement is frequently written as headnote such as in '000 (in thousand) or million (i.e. 10 Lakhs) or Cr. (i.e. crores)

Rules for Tabulation :

1. Captions and stubs should be arranged in systematic order.
2. Measurement should be clearly defined.
3. Avoid overloading the table with details.
4. Table should be logically arranged.
5. Avoid the use of abbreviations.
6. Expression like etc. is not good in table.
7. Dittoes mark should not be used in table.
8. (N.A.) Not Available letter should be use for information not available or even use (dash) can be made for these explanation.
9. (Misc.) Miscellaneous items should be place in the last row of the table.
10. In order to facilitate comparison the arrangement of the categories should be mode according to alphabetical or chorological order.
11. Table must be suitable to the requirements of the study.

Analysis of Data

The data collected may or may not in numerical form. Even if data is not in numerical form still we can carry out qualitative analysis based on the experiences of individual participants.

When data is collected in numerical form than through descriptive statistics findings can be summarised. This includes measure of central tendency like mean range etc. Another way to summarised finding is by means of graphs and charts.

In any of the research study there is experimental hypothesis or null hypothesis. One the basis of data of both hypothesis, various test have been devised to take decision. Where decision is taken on the basis statistical test, it is subject to error, and such correct decision is difficult. But some standard procedures followed to arrived at proper decision.

There is always issue related to the validity of our findings of the study. Where study is not conducted on properly controlled and scientific system, than validity is questionable or the validity may lack when findings of the study cannot be applied to every day life.

In quantitative research information is expressed in numerical form and in qualitative research information is not expressed in numerical form. Quantitative research would measure the number of aggressive acts witnessed. Qualitative research may help to explain why aggressive act occurred.

The cordial principle of qualitative analysis is that causal relationship and theoretical statement emerging out from studies and this theory is not imposed on data.

Analysis involves estimating the values of unknown parameters of the population and testing hypothesis for drawing inferences.

Analysis may be categories as descriptive analysis and inferential analysis.

Descriptive analysis : Is study of distribution of variable. This study may be in respect one, two or move variables. The variation of two variable to decide correlation between two variables can be worked out in correlation analysis.

Casual analysis : Is concerned with how one or move variable affect the charges in another variable.

The functional relationship existing between two variables is termed as regression analysis.

Analysis Procedure :

In order to make effect use of data following tasks are performed by the investigator.

1. An investigator first of all gathers together all the information collected from the respondents.
2. The investigator than arranges the items of information into various groups in a preliminary way.
3. To take account of the categories or grouping suggested by the respondents.
4. The set of categories based on information will be formed.

Then so collected information should he reviewed to assess with reference to objectives of the study. Then data to he analysed with selected tools. The result emerging for this analyses than to be related to the objective of the study.

In qualitative investigation is not only to know the various statements but it is concern with various interpretations the statements in each category. Report in this investigation indicate how hypothesis changed during the course of investigation.

Types of Analysis

1. Qualitative analysis.
2. Content analysis
3. Quantitative analysis.

1. Qualitative Analysis

It is less influenced by theoretical assumptions. The limitation of this type of analysis is that the findings tend to be unreliable. The information categories and interpreted often differ considerable from one investigator to another one.

In this system researcher to go through "research cycle" to increase reliability. Repeating the research cycle is of value in some ways, but it dose not ensure that the findings will have high reliability.

Qualitative analyses are carried out in several different kinds of study like interview, case studies and observational studies

2. Content Analysis

Content analysis is used where originally qualitative information is reduced to numerical terms. It is a method of analysing media output. Which includes articles published in new papers, speeches made in radio, television and various type of propaganda. This method of analysis is applied to all most all from of communications.

The media influences the way we think and feel about issue and so it is important to analyse media communications in detail. Context analysis can revel the issues of concern. The limitation of the context analysis is that it is often very hard to interpret the findings.

3. Quantitative Analysis

The numerical data collected in study through descriptive statistics analysis can be conducted through measure of central tendency. So the descriptive analysis is largely the study of distribution of one variable.

A. Distriptive Analysis

This analysis of one variable is called one dimensional analysis. This analysis measures condition at particular time.

Government of India wanted to study the size of postal savings account as on 31.3.08. The Frequency distribution of the post savings Bank Account is shown in the table given below.

Frequency Distribution of Postal Savings A/c Size

Account Size Rs.	Number of Accounts
Up to 1 - 5000	10,000
50001 - 10,000	8.000
10,000 - 15,000	6,000
15,001 - 20,000	4,000
20.000 - 25,000	2,000
25,000 - 30.000	1,000
30,000 - 35,000	500
35,0001 - 40.000	300
and 40,000 - above	200

This table describe the size of account as on 31.3.08 which further can classify as per the residence or qualification of account holders.

B. Bivariate Analysis

The analysis in respect of two variables is called bivariate analysis. In this analysis collected data in placed into tabular form, so that the meaning of the data can be derived.In this method simple dimensional data is developed and put in to two or more categories. This analysis can be explained with the help of following data of the survey.

A study was conducted tin B grade city to workout their tourism habits. Sample of 1000 persons was randomly interview out of 1000 sample number 350 (35 pc) are regular going on conducted tour and 650 are not regular going on such tours. This fact is shown in following table.

1. Captions and stubs should be arranged in systematic order.

Sample residents tourism habits

Category Number	Percentage	
Regular going on Tours	350	35
Non regular going on Tours	650	65
Total	1000	100

C. Sequential Analysis

When only factor is revel in the table at one time, this type of analysis is called sequential analysis. If we do the further analysis of the same data regard tour going showed that person with leave travel concession (LTC) facilities are more frequently going on tourism than the those who are not gating facilities of L.T.C.

Persons with terns of L.T.C. facilities

Category	Persons with L.T.C.		Persons without L.T.C.		Total
	(Pc)		(Pc)		
Regular going on tours	200	67%	150	21%	350
Non regular going on tours	100	33%	550	79%	650
Total	300	100%	700	100%	1000

The above table revels that the one factor at one time. The person with L.T.C. facilities are more frequently going on conducted tours.

C - Casual Analysis :

It is concerned with study of one variable affecting another one. The dependent variable get affected with external factor. There are various ways to find out the existence on non - existence of coordination's. There analysis is done with the help of statistical tools.

Causal analysis is important in experimental researches.

D. Multivariate Analysis

With an advancement of computer applications there is fast development of multivariate analysis, in which statistical method simultaneously analyse more than two variable on sample of observations. Multiple regression, discriminate analysis and analysis of variance are commonly used for this type of analysis.

E. Inferential Analysis

In order to decide the validity of data to indicate conclusion this analysis is concerned with tests for significance of hypotheses. One the basis of inferential analysis the task of interpretation is performed by estimating the population values.

Statistical Analysis of Data

Statistics is regarded as mathematics applied to, observation of data, the collection, processing, analysis and interpretation of data covered in statistics. It is an application mathematical principles, which help us to understand complex problems in easier way.

The main functions of the statistics is to describe and summarize the information to make it usable and making generalization about population based on sample drawn.

In social research huge data is collected, which requires simplification for its understandability. Now a days with application and quantitative approach it make use of statistical methods more important.

The problem of analysis and interpretation has to be anticipated at every stage in research study. So the knowledge of statistical methods becomes an invaluable tool for social research.

It is also necessary to understand the limitations of statistics, while using statistical techniques is research process. It is normally applied to the problem amenable to quantitative phenomena. Now qualitative data in this form to be converted in quantitative data for statistical analysis.

The statistical laws are applicable to averages of aggregates sometimes the researcher is also tempted to misuses statistics, he may twist data to suit his conclusions. The statistical computations may not be final truth it is an approximation and probabilistic.

In Analysis of data various measure of central tendency, dispersion skew ness, relationship and other measures are considered.

Measures of central tendency describe how the data cluster together around a central point. There are three main measures of central tendency. the mean the median and the mode geometric and harmonic mean.

In measures of dispersion variance and its square root - the standard deviation are most often used as measure.

In measure of skew ness mostly use first measure of skew ness based on mean mode and median. Second measure of skew ness and quartiles.

From among the measure of relationship karl pearson's co-efficient of correlation in case of variables and yules co-efficient of association is used for attributes.

In other measures multiple correlation, co-efficient. partial correlation co-efficient, regression analyses are commonly used. Index numbers, analysis of time series and coefficient of contingency are also used as other measures.

Measures of Central Tendency

A measure if central tendency is a single number used to represent a group of data. It reduces the complexity of the data and makes them easily comparable. Measure of Central Tendency or an Average is a single value which lies at the centre and represents the data. Most of the observations of the data cluster around it. The features or requirements of a good average are as follows.

(a) It should be based on all the observations.

(b) It should be simple to understand.

(c) It should be rigidly defined.

(d) It should not be affected by extreme values.

(e) Lastly it must be useful for further calculations.

Types of Averages (Central tendencies) :

One of the salient characteristics of the distribution of the sample data is that most of the obervations tend to concentrate in the centre of the distribution. This characteristic of distribution in known as Central Tendency.

Central tendency is usually expressed in three ways.

(i) The average value (Arithmetic mean)

(ii) The middle value (median)

(iii) The most frequently occuring value termed as (mode)

Beside these there are two minor measures.

(i) Harmonic mean

(ii) Geometric mean

(i) (Arthmetic mean) ($\bar{x}$) :

The arithmetic mean is the average of all the values of the variate in the sample. If x_1, x_2, x_3, x_n are the n values of the variate x in the sample, then their A. M. is given by,

$$\bar{x} = \frac{x_1 + x_2 + x_3 + \ldots + x_n}{n}$$

$$X = \frac{\varepsilon X}{n}$$

If x, occurs f_1 times, x_2 occurs f_2 times etc..... and lastly x_n occurs f_n times, there being (n) observations altogether.

$$i.e.\ n = f_1 + f_2 + f_3 + \ldots + f_n$$

$$Then\quad \bar{x} = \frac{f_1x_1 + f_2x_2 + f_3x_3 + \ldots + f_nx_n}{f_1 + f_2 + f_3 \ldots + f_n}$$

$$\text{OR}\ \bar{x} = \frac{\varepsilon fx}{\varepsilon f} = \frac{\varepsilon fx}{n}$$

The arthmetic mean is used to report average size, average yield, average percent defictive etc.

The arthmetic mean or simply the mean is the value obtained by dividing the sum of the values of the given items (of a variable) by their number. Thus :

$$\textbf{Mean} = \frac{\text{Aggregate of the values of items in the series}}{\text{Total number of items}}$$

Suppose there are four workers earning a wage of Rs. 12, Rs. 18, Rs. 16, and Rs. 14. The mean wage of these will be obtained by the following question :

$$= \frac{12+18+16+14}{4}$$

$$= \frac{60}{4} = \text{Rs.}\,15$$

In order to sumbolise such computation of mean, if one lets, $x_{1'}$, $x_{2'}$, $x_{3'}, x_n$ etc indicate the various values of the items and n, the number of items x (called'x' -bar), the arithmetic mean derived is,

$$x = \frac{x_1 + x_2 + x_3 \ldots\ldots x_n}{n}$$

Mathematically, the formula, may be abbreviated as :

$$x = \frac{\varepsilon x}{n}$$

Where S ('sigma' - Greek capital letter) denotes the 'sum of,' x represents values of the items. and n stands for the number of items.

The above formula is a direct method of computing a mean.

Computing the mean from a discrete (or ungrouped) frequency distribution

For finding the mean in a discrete frequency distribution, each variate is to be multiplied by its frequency and then aggregate of the products is to be divided by the number of items.

Thus :

$$x = \frac{f_1x_1 + f_2x_2 + \ f_3x_3 + ... + f_nx_n}{f_1 + f_2 + f_3 + ... + f_n}$$

Or simply $x = \frac{\Sigma fx}{n}$

where, f stands for frequency of the variate.

x stands for the variate.

n =ε f , referring to the total number of items.

The following problem will clarify the use of the formula :

Problem No. 1 :

The following data indicate daily earnings (in rupees) of 50 workers in a steel yard :

Daily earnings (in Rs.) :	2	3	4	5	6	7
No workers :	10	12	8	10	5	5

Calculate the average income per worker.

Solution :

Daily Earnings in Rs.	No. of Workers	Product
(x)	(f)	(f×x)
2	10	20
3	12	36
4	8	32
5	10	50
6	5	30
7	5	35
	$\Sigma fx = n = 50$	$\Sigma fx = n = 203$

$$X = \frac{\Sigma fx}{n}$$

$$= \frac{203}{50} = 4.06$$

Answer :

Average income per worker is Rs. 4.06.

Short-cut Method of Computing Mean :

The direct method of computing means as shown above is used only when the items are not very large and their values are also small. If, however, the series is large and/or the values are big, then a short-cut method may be resorted to minimise the calculation task.

In the short-cut method, try to adopt the usual procedure of calculating mean by abbreviating the given data, on the basis of the deviations of the mid-points (m) of class-intervals in a given frequency distribution, from an arbitrary chosen value of (m) as a starting point designated as (a).

In the short-cut method, a is adopted as an arbitrary starting point for computation work. Many statisticians refer a as guess mean or assumed mean.

The following formula is used : $X = a + \frac{\sum fd}{n}$

where,

X = mean;
a = assumed mean;
f = frequencies;
d = $(m - a)$ i.e. deviations of mid-point value from the value of arbitrary origin; and n = $\sum f$ (number of items).

Needlees to say that in the case of discrete distribution, $d = x - a$ where x is the variate.

Problem No. 2 :

Daily wages (Rs.) :		2	3	4	5	6	7
No. of workers :	10	12	8	10	5	5	

solution :

Daily Earnings in Rs.	No. of Workers	Dev. from Arbitrary Assumed Mean : 4	Product
(x)	(f)	$d = (x = 4)$	(fd)
2	10	-2	-20
3	12	-1	-12
4	8	0	0
5	10	+1	+10
6	5	+2	+10
7	5	+3	+15
$n = 50$			$\sum fd = 30$

$$X = a + \frac{\sum fd}{n}$$

$$= 4 + \frac{3}{50}$$

$$= 4.06$$

Problem No. 3 :

Compute the mean wage of the following data

Wage Rs.	No. of Workers
5 - 10	5
10 - 15	25
15 - 20	30
20 - 25	25
25 - 30	10
30 - 35	5

Solution :

Compute the mean wage of the following data

x	f	m	d	fd
5-10	5	7.5	-10	-50
10-15	25	12.5	-5	-125
15-20	30	17.5	0	0
20-25	25	22.5	5	125*
25-30	10	27.5	10	100*
30-35	5	32.5	15	75*
	n = 100			$\sum fd = 125$

$$X = a + \frac{\sum fd}{n}$$

$$= 17.5 + \frac{125}{100}$$

$$= 17.5 + 1.25$$

$$= 18.75$$

Median :

Median is value of item that goes to divide the series into equal parts. It may be defined as the value of that item which divides the series into two equal parts. One half containing values greater than it and the other half containing values less than it. Therefore, the series has to be arranged in Ascending or Descending order before finding the median.

When all the observations are arranged in ascending or descending order, then the median is the magnitude of the middle case, it has half the observation above it and half below it.

If there are 'n' observations of the variate and they are arranged in ascending order, then is given by $\left(\frac{n+1}{2}\right)^{th}$ value if 'n' is odd. On the other hand if 'n' is even the median is taken as the average $\left(\frac{n}{2}\right)^{th}$ of and $\frac{\sigma \ \sqrt{\Sigma x^2} = 274}{= \sqrt{274} = 523}$ value

The median is the second type of central measure tendency, which is the value of item that divides the series into the equal parts.

"The median is that value of the variable which divides the group into two equal parts, one part comprising all values greater and the other, all values less than median."

- By L. R. Connor

"The median as its name indicates, is the value of the middle item in a series, when items are arranged according to magnitude."

- By Yau Lunchow

"The Median is that value which divides a series so that one half or more of the items are equal to or less than it and one half or more of the items are equal to or greater than it."

- By Croxton and Cowden.

Calculation of median :-

Problem No. 1 :-

The following are the marks obtained by 9 students. Find out the median marks.

Marks (out of 100) :- 47, 35, 19, 58, 70, 32, 45, 67, 55

Solution :- Before calculation of median, the above number i.e. marks will be arranged in ascending order and then median will be calculated as follows.

Marks (out of 100)
19
32
35
45
47 (5th item)
55
58
67
70
N = 9

$$M = \text{The size of } \left(\frac{N+1}{2}\right)$$

$$M = \text{The size } \frac{9+1}{2}^{th} \text{ item}$$

$$= \text{The size } \frac{10}{2}^{th} \text{ item}$$

= The size 5th item

Therefore, the median marks = 47

Problem No. 2 :-

Find out the median from the following.

Values : 59, 53, 64, 41, 39, 66, 74, 68

Solution : 39, 41, 53, 59, 64, 66, 68, 74

Therefore n = 8

Values

39

41

53

59 (4th item)

64 (5th item)

66

68

74

N = 8

$$M = \text{The size of} \left(\frac{N+1}{2}\right) th \text{ item}$$

$$= \text{The size of} \left(\frac{8+1}{2}\right) th \text{ item}$$

$$= \text{The size of} \left(\frac{8+1}{2}\right) th \text{ item}$$

= The size of 4.5th item

$$\text{Size of 45th item} = \left(\frac{4^{th}\text{ item} + 5^{th}\text{ item}}{2}\right)$$

$$= \left(\frac{59+64}{2}\right)$$

$$= \left(\frac{123}{2}\right)$$

$$= 61.5$$

Problem No. 3 :

Calculate median value from the following.

Value	4	5	6	7	8	9	10	11	12
Frequency	15	18	20	22	25	30	35	40	50

Value	Frequency	Cumulative frequency (f)
4	15	15
5	18	33 (15+18)
6	20	53 (33 + 20)
7	22	75 (53+22)
8	25	100 (75+25)
9	30	130 (100+30)
10	35	165 (130+35)
11	40	205 (165+40)
12	50	255 (205+50)
	n = 255	

$$M = \text{The size of} \left(\frac{N+1}{2}\right)^{th} \text{item}$$

$$= \text{The size of} \left(\frac{255+1}{2}\right)^{th} \text{item}$$

$= \text{The size of } 128^{th}$ item

= Falls in cf of 130

Hence, median value will be 9.

Problem No. 4 :-

From the following marks distribution calculate the median marks.

Marks (out of 100)	15	20	22	30	35	55	65	70	75
No of students	5	2	8	7	10	8	13	5	2

Solution

Marks (out of 100)	No. of students	C.F.
15	5	5
20	2	7 (5+2)
22	8	15 (7+8)
30	7	22 (15+7)
35	10	32 (22+10)
55	8	40 (32+8)
65	13	53 (40+13)
70	5	58 (53+5)
75	2	60 (58+2)
	n = 60	

$$M = \text{The size of } \left(\frac{N+1}{2}\right) th \text{ item}$$

$$= \text{The size of } \left(\frac{60+1}{2}\right) th \text{ item}$$

$$= \text{The size if } \left(\frac{61}{2}\right) th \text{ item}$$

= The size of 30.5th item

= The falls in C.f of 32

∴ Median Marks will be 35

Problem No. 5 :-

Calculate the median length from the following.

Lenghth in inches	No of units
0-10	2
10-20	15
20-30	28
30-40	50
40-50	70
50-60	45
60-70	35
70-80	10

Solution

Marks (out of 100)	No. of students	(C.f)
1-10	2	2
10-20	15	17
20-30	28	45
30-40	50	95
40-50	70	165
50-60	45	210
60-70	35	245
70-80	10	255
	n = 255	

$$M = \text{The size of } \left(\frac{N}{2}\right) th \text{ item}$$

$$= \text{The size of } \left(\frac{255}{2}\right) th \text{ item}$$

Problem No. 6 :-

The following table shows wages per day and number of workers. Calculate the median

Wages per day	No of workers
Below-50	6
Below-100	16
Below-150	20
Below-200	36
Below-250	40
Below-300	55
Below-350	70
Below-400	100

Solution :-

Wages per day	No. of worker	C.f
0-50	6	6
50-100	10	16 (6+10)
100-150	4	20 (16+4)
150-200	16	36 (20+16)
200-250	4	40 (36+4)
250-300	15	55 (40+15)
300-350	15	70 (55+15)
350-400	30	100 (70+30)
	n = 100	

$$M = \text{The size of } \left(\frac{N}{2}\right) th \text{ item}$$

$$= \text{The size of } \left(\frac{100}{2}\right) th \text{ item}$$

$= \text{The size of } 50^{th} \text{ item}$

= Falls in C.f of 55

∴ Median group will be 250-300

By applying interpolation formula, we get,

$$\text{Median} = l_1 + \frac{l_2 - l_1}{f_1}(m - c)$$

$$\text{Median} = 250 + \frac{300 - 250}{15}(50 - 40)$$

$$= 250 + \frac{50}{15}(10) = 250 + \frac{500}{15}$$

$$= 250 + 33.33 \quad = 283.33$$

Location of Median Graphically

Median can also be calculated graphically with the help of 'oglve' or 'cumulative frequency curves.' The various steps involved in location of median are is follows :

(1) Draw a 'less than ogive.'

(2) Compute N/2 and locate this point in the Y- axis.

(3) Draw a perpendicular from the Y-axis at this point and extend the prependicular to intersect to ogive.

(4) At the intersection, drop a perendicular to the X-axis. The intersection gives the value of median. This is illustrated in Fig. 13.1.

Illustration 28 :

The following table gives the marks obtained by 50 students in Statistics.

Marks	No. of Students
10 marks or less	4
20 marks or less	10
30 marks or less	20
40 marks or less	40
50 marks or less	47
60 marks or less	50

Draw a curve on a graph paper and show the median therein.

Solution : Changing the distribution in an ordinary inclusive class interval series we get :

x	f^1
0 - 10	4
10 - 20	6
20 - 30	20
30 - 40	10
40 - 50	7
50 - 60	5

$$\text{Median} = l_1 + \frac{N/2 - c}{f}(l_2 - l_1)$$

$$= 20 + \frac{25-10}{20}(10)$$

$$= 27.5$$

Graphically the result is shown in Fig. 14.1.

Merits of Median

1. It is easy to calculate and is quickly intelligible.
2. Its value is not much affected by the presence of extreme items.
3. It is the most appropriate average in dealing with data which are not capable of numerical expression at all, e.g. health, intelligence, etc.
4. It can be determined graphically.
5. It is useful in measuring dispersion and skewness.

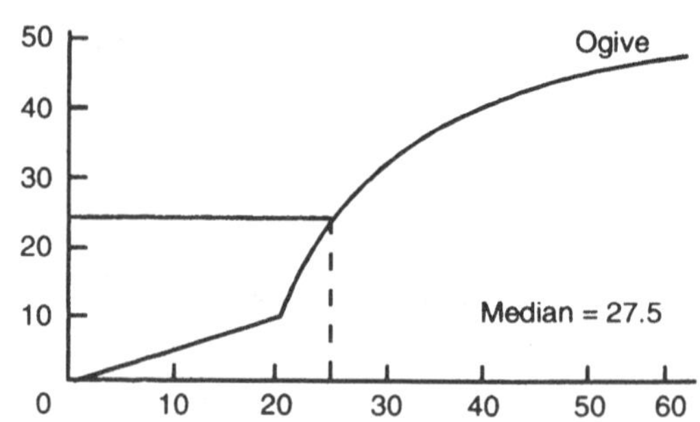

Demerits of Median

1. It requires that data must be arranged in ascending or descending order of magnitude, which involves additional work.
2. It is not open to mathematical manipulations as arithmetic mean is.
3. It is estimated and not computed.
4. If the class intervals are not uniform the value of median become imappropriate.

Mode :

It is the value which occurs with the maximum frequency. It is the most typical or common value that receives the highest frequency. It represents fashion and often it is used in business. Thus, it corresponds to the value of the variable which occurs most frequently. The modal class of a frequency distribution is the class with the highest frequency. It is denoted by a symbol 'Z'.

Mode is the value of variable which is repeated the greatest number of times in the series. It is the usual, and not casual, size of itim in a series. It lies at the position of greatest density. If we say that modal marks obtained by students in a class are 40, we mean that the largest number of students have secured 40 marks. It means around 40 the number of students tend to be heavily concentrated. When we talk of 'most common wage', 'most common height' or 'most common marks', we have in our mind 'Modal Value'. When a doctor makes a statement that the usual duration of cold-fever getting cured is one week, he means that it is the modal duration.

If each observation occurs the same number of times, we can say that there is 'no mode' : If two observations occur the same number of times, we can say that it is a 'Bi-modal' case. If three or more observations occur the same number of times, we can say that it is a 'Multi-modal' case. When there is a single observation which occurs the most number of times, we can say that it is the case of a 'Uni-modal'.

Mode is the value that occurs most frequently, in a frequency histogram or frequency polygon, it is the observed value corresponding to the high point of the graph. However, the mode is not a rigidly defined measure. In fact mode is the most unstable average and its true value is difficult to determine.

For example, the recorded observations are :

(a) 2, 3, 2, 4, 5, 2, 7, 6, 2, 4, 2

Here, 2 occurs most frequently, hence mode = 2

(b) 4, 3, 2, 3, 2, 3, 5, 2, 3, 2

In this case, 2 and 3 both occur equal number of times, hence there are two modes 2 and 3.

The main characteristic of mode is that it is the size of that item which has the maximum frequency and is also affected by neighbouring frequencies.

Calculation of Mode (Z)

Mode is determined by preparing a grouping table and an analysis. A grouping table has the following six columns.

Column 1st : It has original frequencies and the maximum frequency is marked by bold type.

Column 2nd : In this column the frequencies of column 1st are combined two by two. Here also the maximum frequency is marked by bold type.

Column 3rd : Here, we leave the first frequency of column 1st combine the others in two by two. Again the maximum frequency is marked by bold type.

Column 4th : In this column the frequencies of the column 1st are combined (grouped) in three by three and again the maximum frequency is marked by bold type.

Column 5th : Here, we leave the first frequency of the column 1st and group the others in three by three. Again mark the maximum frequency by bold type.

Column 6th : Now leave the first two frequencies of column 1st and combaine the *others* in three by three. Mark the maximum frequence by bold bype.

Problem No.1 :- Determine the mode, from the following daily income of 10 persons.

Daily Income (In Rs.) :- Rs. 500, Rs. 700, Rs. 800, Rs. 650, Rs. 825, Rs. 800, Rs. 700, Rs. 800, Rs. 900, Rs. 1000.

Solution- Rs. 800 repeated three items.

∴ Mode income is Rs. 800. This can be determined by observation only.

Problem No.2 :-

Find out the modal wages from the following.

Wages per day	**No of workers**
100	20
120	25
140	40
160	25
180	30
200	10

Solution :- By observation. Mode can not be determined here the frequency 25 repeated twice in the information. Therefore gropuing of frequencies will be done in six columns, as follows.

Wages	No. of Worker					
	I	II	III	IV	V	VI
100	20	45		**85**		
120	25		**65**		**90**	
140	**40**	**65**				**93**
160	25		55	**65**		
180	30	40				
200	10					

Analysis Table :-

	Wages in Rs.					
Columns	**100**	**120**	**140**	**160**	**180**	**200**
I			I			
II			I	I		
III		I	I			
IV	I	I	I			
V		I	I	I		
VI			I	I	I	
Total	1	3	6	3	1	-

Ans :- It is clear from the above analysis table that modal wages is Rs. 140 because these wages repeated 6 times in the above infomation.

Problem No. 3 : Calculate the mode from the following.

Size of item	frequency
0-10	20
10-20	25
20-30	45
30-40	28
40-50	20
50-60	15
60-70	10

Solution :

Size of item	Frequency					
	I	II	III	IV	V	VI
0-10	20	45		90		
10-20	25		(70)		98	
20-30	(45)	(73)				
30-40	28		48			(93)
40-50	20	35		63		
50-60	15		25		45	
60-70	10					

Analysis Table :-

	Wages in Rs.					
Columns	**0-10**	**10-20**	**20-30**	**30-40**	**40-50**	**50-60**
I			I			
II			I	I		
III		I	I			
IV	I	I	I			
V		I	I	I		
VI			I	I	I	
Total	1	3	6	3	I	-

From the above analysis table, it is clear that group 20-30, repeated 6 times in the information. Therefore modal group will be 20-30. By applying interpolation we will arrive at correct and accurate mode as follows.

$$Z = l_1 + \frac{f_1 - f_0}{2f_1 - f_0 - f_2}(l_2 - l_1)$$

Z = Mode

l_1 = Lower limit of the group

l_2 = Upper limit of the group

f_1 = Frequency of the modal group

f_0 = Frequency of the Preceding group

f_2 = Frequency of the Suceeding group

Putting the values in the formula, we get,

$$Z = L_1 + \frac{f_2 - f_0}{2f_1 - f_0 - f_2}(L_2 - L_1)$$

$$= 20 + \frac{45 - 25}{2(45) - 25 - 28}(30 - 20)$$

$$= 20 + \frac{20}{90 - 25 - 28}(10)$$

$$= 20 + \frac{200}{37}$$

$= 20+5.41$

$= 25.41$

Problem No. 4 :-

Calculate the middle value of marks obtained from the following.

Marks obtained (out of 100)	No of Students
10-19	5
20-29	30
30-39	88
40-49	180
50-59	250
60-69	200
70-79	130
80-89	40
90-99	10

Solution :- The above inclusive continuous series must be converted into the exclusive continous series, by deducting 0.50 from lower limit of a group and adding 0.5 in upper limit because there is difference of 1 in upper limit of previous group and lower limit of next group.

Marks obtained	No. of Students					
	I	II	III	IV	V	VI
9.5-19.5	5	35				
19.5-29.5	30			123		
29.5-39.5	88		118		298	
39.5-49.5	180	268				
49.5-59.5	250		430	630		518
59.5-69.5	200					
69.5-79.5	130	450	330		580	
79.5-89.5	40			180		370
89.5-99.5	10	170	50			

Analysis Table

Columns	**9.5 - 19.5**	**19.5 -29.5**	**29.5 -39.5**	**39.5 - 49.5**	**49.5 -59.5**	**59.5 -69.5**	**69.5 -79.5**	**79.5 - 89.5**	**89.5 99.5**
I					I				
II						I	I		
III				I	I				
IV	I	I	I						
V					I	I	I		
VI			I	I	I				
Total	1	1	2	2	4	2	2	-	-

From the above, it is clear that group 49.5-59.5 repeated six times, Hence modal group will be 49.5-59.5

By applying the interpolation formula, we get

$$Z = l_1 + \frac{f_1 - f_0}{2f_1 - f_0 - f_2}(l_2 - l_1)$$

Putting the values in the formula, we get,

$$Z = 49.5 + \frac{250 - 180}{2(250) - 180 - 200}(59.5 - 49.5)$$

$$= 49.5 + \frac{70}{500 - 180 - 200}(10)$$

$$= 49.5 + \frac{700}{120}$$

$$= 49.5 + 5.83$$

$$= 55.3$$

Merits and Demetits of Mode

Merits :

1. It is widely used in production and trade.
2. It is easy to understand and can be located even by inspection.
3. All the items are not needed in estimating Mode.
4. It unaffected by the extreme items.
5. It can be located graphically.
6. It represents the distribution in a better manner.

Demerits :

1. Mathematical manipulation is not possible.
2. It is uncertain and indefinite.
3. There can be more than one mode in a distribution.

Relationship between Mean, Median and Mode

Mean, median and mode have their distnct role in statistical analysis. In no case they can be substitited for one another. Int a moderately asymetrical distributiob, the following relationship exists.

In Fig. 14.2 OZ = mode, OM = median and OX = arithmetic mean. It is also clear in Fig. 14.2 that

2MX = ZM and #MX = ZX

where MX = Mean - Median

ZM = Median - Mode

ZX = Mean - Mode

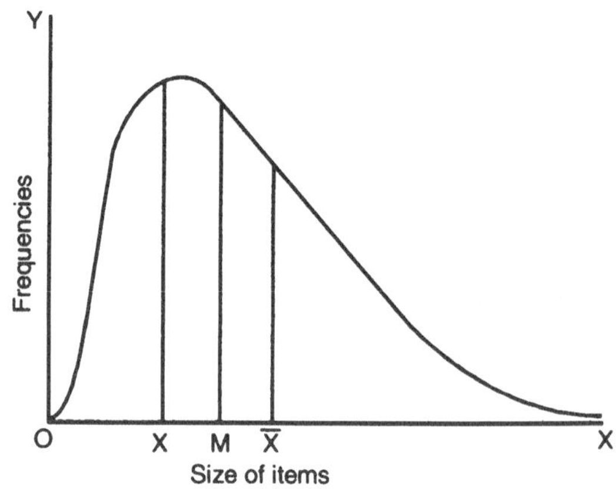

3 MX = ZX provides us the following relationship between three measures of central tendency.

3 (Mean = Median) = Mean - Mode

3 Mean - 3 Median = Mean - Mode

or Mode = 3Median - 2 Mean ...(i)

Using eq. (i) we get

Mode = 3 Median - 2 Mean

2 Mean = 3 Median - Mode

Mean = : (3 Median - Mode) ...(ii)

or 3 Median = Mode + 2Mean)

Median = (Mode + 2Mean) ...(iii)

These three equation are vital for knowing the reletionship between mean, median and mode are 18.8 and 20.2 respectively. Calculate the value of mode.

Solution : We know :

Mode

= 3 Median = 2 Mean

= 3 x 18.8-2 x 20.2 = 56.4 x 40.4 = 16.0

Measures of Skewness

Another measure for the study of a distribution is known as skewness. Measures of central tendency tell us about the central value of a distribution; measures of variability tell us about the concentration or dispersion of items from the central value. Measures of skewness tell us whether the dispersion of items from and average is symmetrical or asymmetrical.

When the dispersal of items from an average is symmetrical, i.e., number of items below an average equal the number of items above it, we call the distribution normal distribution. In a normal distribution the value of

Mean = Median = Mode

If a distribution is not normal distribution, it is call skewed distribution, i.e., a distribution in which skewness exists. In a skewed distribution, values of mean, median and mode do not coincide.

Skewness may take two forms : positive skewness and negative skewness.

In case of a positive skewness,

Mean > Median > Mode

In case of negative skewness,

Mean < Median < Mode

A normal distribution when plotted on a graph, gives us a normal curve. A normal curve is a Bell-shaped curve.

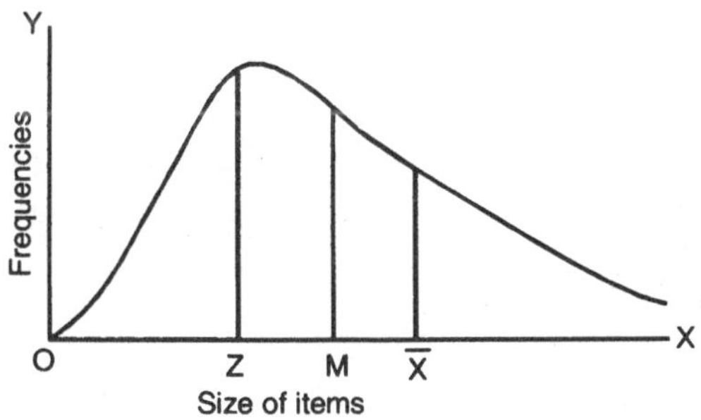

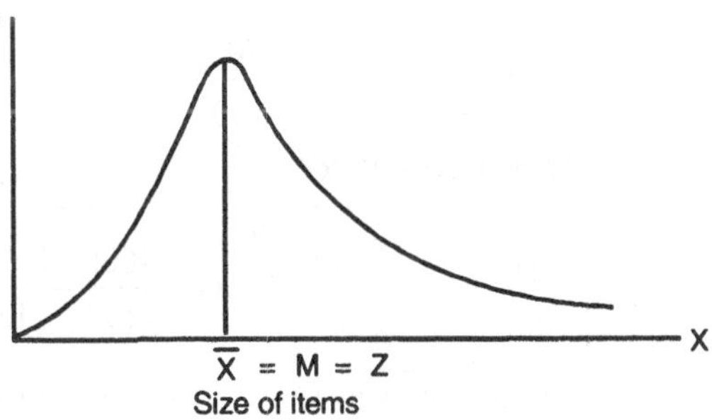

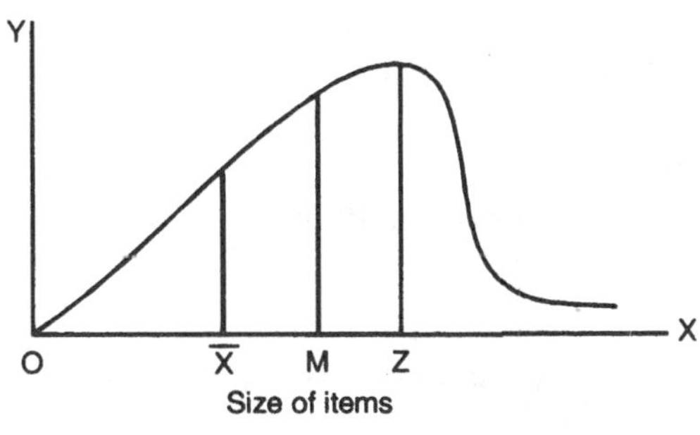

The properties of a normal curve are as follows :

1. The values of mean, median and mode are indentical.
2. Pairs of quartilies (i.e., q_a and q_1) are equidistant from median.
3. Sums of positive and negative deviations from median are equal.
4. Number of frequencies below and above mode are equal.

In a skewed curve, on the other hand, the tail tapers off to the left, A normal curve shows absence of skewness.

Problem 1

Calculate Karl Pearson's Coefficient of Skewness from the following data :

Marks above	:	0	10	20	30	40	50	60	70	80
No of students	:	150	140	100	80	80	70	30	14	0

Solution :

Changing the cumulative frequency distribution into exclusive class-interval series.

	f	M.V.	dx	dx'	fdx'	fdx'^2	c.f.
0 - 10	10	5	-30	-3	-30	90	10
10 - 20	40	15	-20	-2	-80	160	50
20 - 30	20	25	-10	-1	-20	20	70
30 - 40	0	35	0	0	0	0	70
40 - 50	10	45	10	1	10	10	80
50 - 60	40	55	20	2	80	160	120
60 - 70	16	65	30	3	48	144	136
70 - 80	14	75	40	4	56	224	150
Ef = 150					Efdx' = 64		fEdx'2 = 808

$$Y = a + \frac{Efdx'}{Ef} \times i = 35 + \frac{64}{150} \times 10$$

$= 35 + 4.26 = 9.26$

$$\sigma = \sqrt{\frac{Efdx'^2}{Ef} - \left(\frac{Efdx'}{Ef}\right)^2} \; XI = \sqrt{\frac{808}{150} - \left(\frac{64}{150}\right)^2} \times 10$$

$$= \sqrt{5.387 - 0.182} \times 10 = \sqrt{5.205} \times 10 = 22.814$$

$$\text{Median} = \text{Size of} \left(\frac{N}{2}\right) \text{item} = \text{Size of} \left(\frac{150}{2}\right) \text{item}$$

= Size of 75th item

The median item lies in the group of 40-50.

By the formula of interpolation.

$$\text{Median} = l_1 + \frac{(N/2) - c}{f}(l_2 - l_1)\,40 + \frac{75 - 70}{10}(50 + 40) = 45$$

Since the distribution is biomodal so we shall apply Karl Pearson's second formula.

$$f = \frac{3(\text{Mean} - \text{Median})}{a} = 3\frac{(39.26 - 45)}{22.8}$$

$$= \frac{3\times\ 5.74}{22.8} = \frac{17.22}{22.8} = -0.755.$$

There is a negative skewness.

Problem 2 :

Find out Karl Pearson's coefficient of skewness from the data given below :

Wages Rs.	:	70-80	80-90	90-100	100-110
No. of workers	:	12	8	35	42
Wages Rs.	:	110-120	120-130	130-140	140-145
No. of workers	:	50	45	20	8

Solution :

X	M.V.	dx	dx'	*f*	*f*dx'	*f*dx'^{2}
70 - 80	80	-40	-4	12	-48	192
80 - 90	90	30	-3	18	-54	162
90 - 100	95	-20	-2	35	-70	140
100 - 110	105	-10	-1	42	-42	42
110 - 120	115	0	0	50	0	0
120 - 130	125	10	1	45	45	45
130 - 140	135	20	2	20	40	80
140 - 150	145	30	3	8	24	72
				Ef = 230	Efdx' = -105	Efdx'^{2} = 733

$$X = a + \frac{}{Ef} \times i = 115 - \frac{105}{230}\times 10$$

$$= 115 - 4.57 = 110.43$$

The mode lies in the group of 110-120

By the formula of interpolation

$$\text{Mode} = 11 + \frac{f_m - f_1}{2f_m - f_1 - f_2}(l_2 - l_1)$$

$$= 110 + \frac{50 - 42}{100 - 45 - 42}(120 - 110) = 110 + \frac{8\times 10}{13} = 110 + 6.15$$

$$= 116.15$$

$$\sigma = \sqrt{\frac{Efdx'^2}{Ef}\frac{Efdx'}{(Ef)}} \times i = \sqrt{\frac{733}{230}\frac{105}{(230)^2}} \times 10$$

$\sigma = \sqrt{3.1869 - 0.2084} \times 10 = \sqrt{2.9785} \times 10 = 1.725 \text{ x } 10 = 17.25$

Karl Pearson's Coefficient of Skewness

$$f = \frac{(\text{Mean} - \text{Mode})}{a} = \frac{110.43 - 116.15}{17.25} = \frac{-5.72}{17.3} = -0.33$$

The coefficient of Skewness is negative 0.33.

Problem 3

Consider the following distribution :

	Distribution A	*Distribution B*
Mean	100	90
Median	90	80
Standard deviation	10	10

(i) Distribution A has the same degree of variation as Distribution B. Do you agree? Give reasons.

(ii) Both distribution have the same degree of skewness. True/False? Give reasons.

Solution : Coefficient of variance

In Distribution $A = \frac{a}{X} \times 100 \text{ or } \frac{10}{100} \times 100 = 10$

In Distribution $B = \frac{10}{90} \times 100 = 11.1$

The degree if variation in Distribution B is more as the value of coefficient of variance in this case is more.

(ii) Coefficient of Skewness

Since Coefficient of skewness $= \frac{3(\text{Mean} - \text{Median})}{a}$

In Distribution A = $A = \frac{3(100 - 90)}{10} = 3$

In Distribution B = $B = \frac{3(90 - 90)}{10} = 3$

Both the distrinutions have same degree of skewness, as the standard deviation and the difference between mean and median is also same.

Illstration 4 : For a moderately skewed data, the arithmetic mean is 100, the coefficent of variation is 35, and Karl Pearson's coefficient of skewness is o.2 Find the value of mode and median.

Solution : In this question the following results are given :

X = 100; C.V.= 35 f = 0.2

Since C.V.= $= \frac{a}{X} \times 100 \text{ or } 35 = \frac{a}{100} \times 120 a = 35$

$$J = \frac{\text{Mean} - \text{Mode}}{a}$$

$$0.2 = \frac{100 - \text{Mode}}{35} \text{ or } 7.0 = 100 - \text{Mode, or Mode} = 93$$

Since in a moderately skewed distribution
Mean - Mode = 3 (Meadian-Median) or 7 = 300 - 3 Median
3 Median = 293 or Median = 7.663

Geometric Mean :

The geometric mean (G.M.) of a set of 'n' observations is the nth root of their product. Hence, if X_1, X_2... X_n are the given 'n' observarions, then

$$\text{G.M.} = \sqrt{(x_1) \times (x_2) \times \ldots\ldots.(x_n)}$$

Thus, the G.M. of 3 values 2, 3 4 would be

$$\text{G.M.} = \sqrt[3]{2 \times 3 \times 4} = 2.885$$

If n = 2 *i.e.*, two observations, then G.M. can be computed by taking the square root of their product.

$$\text{G.M. of 4 and 16 is } \sqrt{4 \times 16} = \sqrt{64} = 8$$

When the number of obseavations is three or more, then determination of *G.M.* is difficult. Hence, to simplify,. logrithmis are used. It is as follows :

$$\log \text{G.M.} = \frac{\log X_1 + \log X_2 + \log X_3 + \ldots X \log X_n}{n}$$

$$\text{Alternatively, } \log \text{G.M.} = \left(\frac{\Sigma \log X}{n}\right)$$

$$\therefore \quad \text{G.M.} = \text{Antilog}\left(\frac{\Sigma \log X}{n}\right)$$

Formula, therefore, in individual series is

$$\text{G.M.} = \text{Antilog}\left(\frac{\Sigma \log X}{n}\right)$$

In case of Ungrouped data/Discrete series

$$\text{G.M.} = \text{Antilog}\left(\frac{\Sigma f \log X}{n}\right)$$

$$\text{G.M.} = \text{Antilog}\left(\frac{\Sigma\ f \log m}{n}\right)$$

Steps involved :

(a) Take the log of the variable X and then total it, i.e. $\Sigma \log X$.

(b) Now, divide $\Sigma \log X$ by 'n' observations and take the antilog of the value so obtained.

Problem 1:

Calculate G.M. from the following data :

(X)	($\log X$)
125	2.969
1462	3.1650
38	1.5798
7	0.8451
0.22	$\bar{1}.3424$
0.08	$\bar{2}.9031$
12.75	1.1055
0.5	$\bar{1}.6990$
$\Sigma \log X$ = 6.7368	

$$\text{G.M. Antilog} \left(\frac{\Sigma \log X}{n}\right) = \left(\frac{6.7368}{8}\right)$$

$$\text{Antilog} = (0.8421) = 6.952$$

(2) (Ungrouped Data) :

(a) Find the log of variable X.

(b) Now, product these log with the repective frequencies.

(c) Total $\Sigma \log X$

(d) Divide the value obtained with 'n' observation (n = ε f)

(e) Take antilog of the value obtained

$$\text{G.M. Antilog } \frac{\varepsilon F \log x}{n}$$

(3) (Grouped Data) :

$$\text{G.M.} = \text{Antilog } \frac{\Sigma f \log m}{n}$$

Steps involved :

(a) First find the mid-points and take their log.

(b) Product these logathirms with respective frequencies.

(c) Total $(\Sigma f \log m)$

(d) Divide the total obtained by 'n' (n = Σf).

(e) Take the antilog of the value so obtained.

Problem 2 :

Find the G.M. of the data given below :

Solution :

Marks (X)	Frequency (f)	m	log m	f x log m
4-8	6	6	0.7782	4.6692
8-12	10	10	1.0000	10.0000
12-16	18	14	1.1461	20.6298
16-20	30	18	1.2553	37.6590
20-24	15	22	1.3424	20.1360
24-28	12	26	1.4150	16.9800
28.32	10	30	1.4771	14.7710
32.36	6	34	1.5315	9.1890
36-40	2	38	1.5798	3.1596
	$\Sigma f = 109$		$\Sigma f \times$ log	$m = 137.1936$

$$\text{G.M.} = \text{Antilog } \frac{\Sigma f \log m}{n} = \frac{137.1936}{109}$$

$$\text{Antilog} = 1.2587$$
$$= 18.14$$

Advantages of G.M. :

(i) Like Arthmetic Mean, G.M. is based on all items.

(ii) It is capable of further algebraic treatment.

(iii) It has great utility in average rates percentages and in determining rates of increase and decrease.

(iv) G.M. gives more weight to small items and less to large items.

Harmonic Mean :

Harmonic Mean of a series is the reciprocal of the arithmetic average of the reciprocals of the values of its various items.

Symbolically the Harmonic mean or HM if a series is

$$\text{HM} = \text{Reci} = \frac{\frac{1}{x_1} + \frac{1}{x_2} + \ldots\ldots\ldots \frac{1}{x_n}}{N} \text{ where}$$

X stands for the value of the variable.

Thus the Harmonic Mean of 2, 4 and 8 would be reciprocal $\frac{\frac{1}{2}+\frac{1}{4}+\frac{1}{8}}{3}$ or reciprocal of $\frac{7/8}{3}$ or reciprocal of 7/24 or 24/7 = 3.43

If the number of items in a series is large it would be a tedious job to find the reciprocal of each items and then to total them and then to divide the total by the number of items and then to find out the reciprocal of the value. The formula can be simplified in the following manner.

$$\text{H.M.} = \text{Reci.} \frac{\frac{1}{x_1} + \frac{1}{x_2} + \ldots\ldots\ldots \frac{1}{x_n}}{N} \quad \text{or}$$

$$\text{HM} = \text{Reci.} \frac{N}{\frac{1}{x_1} + \frac{1}{x_2} + \ldots\ldots\ldots \frac{1}{x_n}} \quad \text{or}$$

$$= \frac{N}{\Sigma\ (1/X)} \ldots\ldots$$

Thus H.M. is equal to the number if items of N divided by the sum of the reciprocals of different items.

This formula is used in the series of individual observations.

In discrete series :

We first find out the reciprocal of each value and multiply it by the concerned frequency. Then total the products and divide the total by the total frequency and then find out ther reciprocal of the value, which is the harmonic mean of the series. Thus in a discrete series :

H.M. Reciprocal of -

$$\text{H.M.} = \frac{f_1 \left(\frac{1}{X_1}\right) + f_2 \left(\frac{1}{X_2}\right) + \ldots\ldots\ f_n \left(\frac{1}{X_n}\right)}{\Sigma f}$$

This formula can agrin be simplifed as

$$\text{H.M.} = \frac{\Sigma f}{\Sigma \left(f \times \frac{1}{\text{X}}\right)} = \frac{\text{N}}{\Sigma \left(f \times \frac{1}{\text{X}}\right)}$$

In continuous series, the value of the variable, is the mid-point of the less-interval or m.v. = X and the formula for the calculation of H.M. in such a series would be

$$\text{H.M.} = \frac{\Sigma f}{\Sigma \left(f \times \frac{1}{m}\right)} = \frac{N}{\Sigma \left(f \times \frac{1}{m}\right)}$$

We shall now use these formulae to calculate the H.M. in the following examples.

Calculation of Harmonic in a series of individual observations.

Problem 1: Calculate the Harmonic mean of the following values :

15, 250, 15.7, 157, 1.57, 105.7, 10.5 1.06, 25.7 and 0.257.

Solution :

We shall find out the reciprocals of the above vakues from the mathematical tables which are available, instead of manual calculation.

Calculation of Harmonic Mean

Values (X)	Reciprocal (1/X)
15	0.06667
250	0.00400
15.7	0.06369
157	0.00637
1.57	0.63690
105.7	0.00946
10.5	0.09524
1.06	0.94340
25.7	0.03891
0.257	3.89100
N=10	Σ (1/X) = 5.75564

$$\text{H.M. Rec.}\left(\frac{\sum \text{Reci } X}{N}\right)$$

$$= \text{Reci } \frac{5.75564}{10}$$

$$= \text{Reci } 0.575564$$

$$= 1.737$$

However a simpler procedure would be

$$\text{HM} = \frac{N}{\varepsilon(1/x)} = \frac{10}{5.75564}$$

$$= 1.737$$

Problem No. 2

Calculate the Harmonic Mean of the following series :

Value :	2	6	10	14	18
Frequency :	4	12	20	9	5

Solution :

Calculation of Harmonic Mean

Values (X)	Frequency (f)	$(f \times \frac{1}{X})$ or $\left(\frac{f}{X}\right)$
2	4	2.000
6	12	2.000
10	20	2.000
14	9	0.6429
18	5	0.2778
	N = 50	$\Sigma \left(\frac{f}{X}\right)$ =6.9207

$$\text{H.M.} = \frac{N}{\Sigma\left(\frac{f}{x}\right)}$$

$$= \frac{50}{6.9207}$$

$$= 7.25$$

We have used the simplified formula for the calculation of harmonic mean, otherwise we would have to find the reciprocal of each value and multiply it by the respective frequency and then total the products and then divide it by the number of items. It would then have been a tedious process.

Index Number

If we have some money, we can purchase goods or services. But it is not always possible to buy the same quantity of goods or services out of that amount, because the pruchasing power of money i.e. value of money changes frequently, some times money has greater purchasing power, sometimes less at different periods. The change in money value cannot be measured. It is related to price level. The value of money decreases or increases measured on the basis of price increases or decreases.

An index number is a number, which is used to measure the level of a certain commodity as compared to the level of the same commodity at some standard period. An Index Number is a statistical instrument for comparing the general level magintude of a group of related variables in two or more situations. Whenever a comparison has to be made the index number are used. To day, for comparative study of price level or changes in consumption, export, import, cost of living, standard of living and national income etc. are studied with the help of index numbers.

Limitations of Index Number

(1) Index numbers, if they are not properly constructed will not be useful as economic barometer.

(2) Biased persons my select a suitable year for themselves as the base year to get desired results.

(3) There can be error in the construction of the index number like in selection of commodities, selection of the base period etc.

(4) Because index numbers are based on sample data. They can represent the exact change in price level.

(5) There is no formula for measuring the price change. There is the formula error in each index. Therefore it will not be representative data.

(6) The weight of data may not be suitable for present data, as tastes, habits and custome of people change in course of time.

Notations of Index Numbers :

Base Year : The year selected for which comparisons are made. It is denoted as 'o'

Current Year : The years for which comparisons are required. It is denoted as 'I'

P_0 = Price of commodity in the base year.

P_1 = Price of commodity in the current year.

q_0 = Quantity of commodity consumed or purchased during the base year.

q_1 = Quantity of commodity consumed or purchased during the current year.

p_{01} = Price index of current year on the basis of base year's information.

Weighted Aggregate Price Index Numbers :

There are various methods of putting weights and a large number of formulae for constructing index numbers have been developed. Some of the important formulas are as follows.

(1) Laspeyr's Method :- Laspeyres price index is based on the assumption that the quantities consumed in the base year and the current year are the same. The price change would not influence the quantity demanded. He has given the following formula.

$$P_{01} = \frac{\Sigma P_1 q_0}{\Sigma P_0 q_0} \times 100$$

(2) Paasche's Method :- In this method, quantities of current year are used as weights. The two indexes give different results when applied to the same data. The relationship between the two indexes depends on the correlation between the price and quantity movements. Paasche's. Index formula is as follows.

$$\frac{\Sigma P_1 Q_1}{\Sigma P0Q_1} \times 100$$

(3) Bowley and Dorbish's Method :- In this method index number is calculated as arthmetic mean of Laspeyr's and Paasche's methods. It is represented as follows.

$$P_{01} = \frac{\frac{\Sigma P_1 q_0}{\Sigma P_0 q_0} + \frac{\Sigma P_1 q_1}{\Sigma P_0 q_1}}{2} \times 100$$

(4) Fisher's Ideal Method :- Fisher has given the formula based on geometric mean of Laspeyr's and Paasche's formula. Fisher's ideal formula of index number is as follows.

$$P_{01} = \sqrt{\frac{\Sigma P_1 Q_0}{\Sigma P_0 Q_0} \times \frac{\Sigma P_1 Q_1}{\Sigma P_0 Q_1}} \times 100$$

Problem 1

Construct index number of price from the following data as per

(1) Laspeyr's Method (2) Paasche's Method

(3) Bowley's Method (4) Fisher's ideal formula

Commodities	Base Year		Current year	
	kgs. Quantity	Rate in Rs Price	kgs. Quantity	Rate in Rs. Price
Bread	10	3	8	3.25
Butter	20	15	15	25
Milk	5	20	8	22

Solution

Commodities	Base Year		Current year					
	Qty. Price P_0(1)	Qty.in in Kgs. Q_0(2)	Price P_1(3)	P_0Q_0 Kgs. Q_1(4)	P_1Q_1 (1x2)	P_0Q_1 (3x4)	P_1Q_0 (1x4)	
Bread	3	10	3.25	8	30	26	24	32.50
Butter	15	20	25	15	300	375	225	500.00
Milk	20	5	22	8	100	176	160	110.00
					$\Sigma P_0Q_0 = 430$	$\Sigma P_1Q_1 = 557$	$\Sigma P_0Q_1 = 490$	$\Sigma P_1Q_0 = 642.5$

Calculation of index numbers :-

(1) Laspeyr's Method -

$$P_{01} = \frac{\Sigma P_1 q_0}{\Sigma P_0 q_0} \times 100$$

$$= \frac{642.50}{430} \times 100$$

$$= 149.42$$

(2) Passche's Method -

$$P_{01} = \frac{\Sigma P_1 Q_1}{\Sigma P_0 Q_1} \times 100$$

$$P_{01} = \frac{577}{409} \times 100$$

$$= 141.08$$

(3) Bowley's Method -

$$P_{01} = \frac{\frac{\Sigma P_1 Q_0}{\Sigma P_0 Q_0} + \frac{\Sigma P_1 Q_0}{\Sigma P_0 Q_0}}{2} \text{ X } 100$$

$$P_{01} = \frac{\frac{642.50}{430} + \frac{577}{409}}{2} \times 100$$

$$= \frac{1.49 + 1.41}{2} \times 100$$

$$= \frac{2.90}{2} \times 100$$

$$= 1.45 \times 100$$

$$= 145$$

(4) Fisher's Ideal index Method :

$$P_{01} = \sqrt{\frac{\Sigma P_1Q_0}{\Sigma P_0Q_0} + \frac{\Sigma P_1Q_1}{\Sigma P_0Q_1}} \times 100$$

$$P_{01} = \sqrt{1.49 \times 1.41} \times 100$$

$$P_{01} = \sqrt{2.10} \times 100$$

$$P_{01} = 1.449 \times 100$$

$$P_{01} = 144.9$$

Problem 2

Calculate the Index Numbers for the year 2001, by taking 1991 as base year as per (1) Laspeyr's Method, (2) Paasche's Method

(3) Bowley's Method, (4) Fisher's Ideal index number.

Commodities	Base Year (1991)		Current year (2001)	
	Price in Rs.	Quantity	Price in Rs.	Quantity
Wheat	500	10	800	12
Sugar	800	15	1000	18
Rice	1000	5	1200	7
Dal	700	12	1000	15

Solution

Commodities	Base Year (1991)		Current year (2001)		P_0Q_0	P_1Q_1	P_0Q_1	P_1Q_0
	Price P_0(1)	Qty. Q_0(2)	Price P_1(3)	Qty Q_1(4)	(1x2)	(3x4)	(1x4)	(3x2)
Wheat	Rs. 500	10	Rs. 800	12	5000	9600	6000	8000
Sugar	Rs. 800	15	Rs.1000	18	12000	18000	14400	15000
Rice	Rs.1000	5	Rs.1200	7	5000	8400	7000	6000
Dal	Rs. 700	12	Rs.1000	15	8400	15000	10500	12000
					30400	51000	37900	41000

Calculation of Index Numbers

(1) Laspeyr's Method -

$$P_{01} = \frac{\Sigma P_1Q_0}{\Sigma P_0Q_0} \times 100$$

$$P_{01} = \frac{41000}{30400} \times 100$$

$$P_{01} = 1.3487 \times 100$$

$$P_{01} = 134.87$$

(2) Paasche's Method

$$P_{01} = \frac{\sum P_1Q_1}{\sum P_0Q_1} \times 100$$

$$P_{01} = \frac{51000}{37900} \times 100$$

$$P_{01} = 1.3456 \times 100$$

$$P_{01} = 134.56$$

(3) Bowely's Method -

$$P_{01} = \frac{\frac{\sum P_1Q_0}{\sum P_0Q_0} + \frac{\sum P_1Q_1}{\sum P_0Q_1}}{2} \times 100$$

$$P_{01} = \frac{\frac{41000}{30400} + \frac{51000}{37900}}{2} \times 100$$

$$= \frac{1.3487 + 1.3456}{2} \times 100$$

$$P_{01} = \frac{2.6943}{2} \times 100$$

$$P_{01} = 1.34715 \times 100$$

$$P_{01} = 134.715$$

(4) Fisher's Ideal index Method -

$$P_{01} = \sqrt{\frac{\sum P_1Q_0}{\sum P_0Q_0} \times \frac{\sum P_1Q_1}{\sum P_0Q_1}} \times 100$$

$$p_{01} = \sqrt{1.3487 + 1.3456} \times 100$$

$$P_{01} = \sqrt{\frac{41000}{30400} \times \frac{51000}{37900}} \times 100$$

$$P_{01} = \sqrt{1.8148} \times 100$$

$$P_{01} = 1.34714 \times 100$$

$$P_{01} = 134.714$$

Problem 3

find out Index Numbers by following Methods

1. Laspeyr's Method,
2. Paasche's Method,
3. Bowley's Method,
4. Fisher's ideal index number

Particulars	Years	X	Y	Z
Price per unit	1995	Rs. 60	Rs. 80	Rs. 10
Quantity	2000	100 kgs.	150 kgs.	5 kgs.
Quantity	1995	50 kgs.	25 kgs.	10 kgs.
Price per unit	2000	Rs. 50	Rs. 40	Rs. 8

Solution

Commodities	**Base Year (1995)**		**Current year (2001)**		**P_0Q_0 (1x2)**	**P_1Q_1 (3x4)**	**P_0Q_1 (1x4)**	**P_1Q_0 (3x2)**
	Price P_0(1)	**Qty. Q_0(2)**	**Price P_1 (3)**	**Qty Q_1 (4)**				
X	Rs.60	50kgs	Rs.50	100kgs	3000	5000	6000	2500
Y	Rs.80	25kgs	Rs.40	150kgs	2000	6000	1200	1000
Z	Rs.10	10kgs	Rs.8	5kgs	100	40	50	80
					ΣP_0Q_0=5100	$\Sigma P_1Q_1$1104	ΣP_0Q_1=18,050	ΣP_1Q_0=3580

(1) Laspeyr's Method

$$P_{01} = \frac{\Sigma P_1Q_0}{\Sigma P_0Q_0} \times 100$$

$$P_{01} = \frac{3580}{5100} \times 100$$

$$P_{01} = 70196 \times 100$$

$$P_{01} = 70.196$$

(2) Passche's Method

$$P_{01} = \frac{\Sigma P_1Q_1}{\Sigma P_0Q_1} \times 100$$

$$P_{01} = \frac{11040}{18050} \times 100$$

$$P_{01} = 0.61163 \times 100$$

$$P_{01} = 61.1634$$

(3) Bowley's Method

$$P_{01} = \frac{\frac{\Sigma P_1Q_0}{\Sigma P_0Q_0} + \frac{\Sigma P_1Q_1}{\Sigma P_0Q_1}}{2} \times 100$$

$$P_{01} = \frac{\frac{3580}{5100} + \frac{11040}{18050}}{2} \times 100$$

$$P_{01} = \frac{0.70196 + 0.61163}{2} \times 100$$

$$P_{01} = \frac{1.31359}{2} \times 100$$

$$= 0.656795 \times 100$$

$$= 65.6795$$

(4) Fisher's Ideal index Method

$$P_{01} = \sqrt{\frac{\Sigma P_1Q_0}{\Sigma P_0Q_0} + \frac{\Sigma P_1Q_1}{\Sigma P_0Q_1}} \times 100$$

$$P_{01} = \sqrt{\frac{3580}{5100} + \frac{11040}{18050}} \times 100$$

$$P_{01} = \sqrt{0.70196 \times 0.61163} \times 100$$

$$P_{01} = \sqrt{0.4293} \times 100$$

$$P_{01} = 0.6552 \times 100$$

$$P_{01} = 65.52$$

Analysis of Time Series

The analysis of such figures is called the Analysis of Time series. Time series analysis is done primarily for the purpose of making forecasts for future and also for the purpose of evaluating past performances. Forecasting for future done by analysing the past behaviour of the variable under study. Hence , the analysis of time series assumes as great importance in the study of all researchers problems.

Components of a Time Series

The changes in the values of a variable related to time can be the result of a large variety of factors, like changes in the tastes and habits of people, changes in population, reduction of cost of production, increase in incomes of people etc. The value of a variable changes due to the interaction of such forces.

If these forces were constant of not liable to charge the value of the variable would also be constant of even if an equilubrium of their effects was slightly disturbed and after this there was no change, the values of the variable would also slightly change and then become constant.

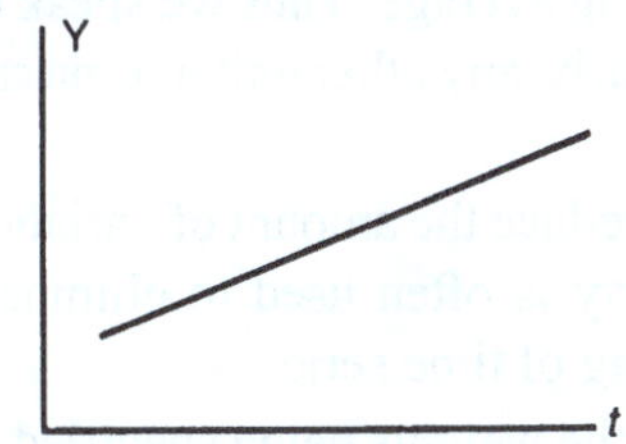

(A) Long-Term Trend

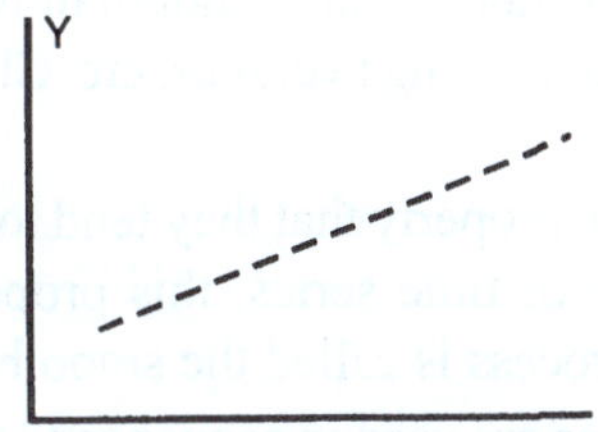

(B) Long-Term Trend and Cyclical Movement

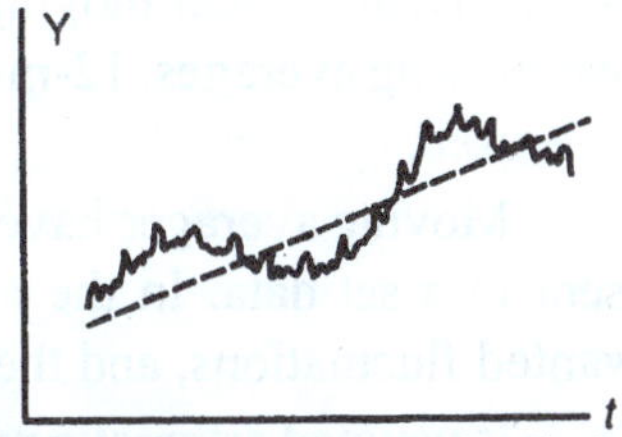

(C) Long-Term Trend , Cyclical and Seasonal Movement

ideal time series. Figure (A) shows the graph of long-term, or secular, trend line (rather than a trend curve, which could have been used instead), Fig (B) shows this long-term trend line with a superimposed cyclic movement (assumed to be periodic), and Fig. (C) shows a seasonal movement superimoposed on Fig (B). If we were to superimpose some irregular, or random, movements on Fig. (C), the result would look more like the actual time seris that occur in practice.

The concepts illustrated in Fig. (B) suggest a technique for analyzing time series. We assume that the time-series variable Y is a product of the variables T, C, S, and *I* that produce the trend, cyclic, seasonal and irregular movements, respectively. In symbols,

$$Y = T \times C \times S \times I = TCSI$$

Time-series analysis amounts to investigating the factors *T, C, S,* and *I* and is often referred to as a decomposition of a time series into its basic component movements.

It should be mentioned that some statisticians prefer to consider Y to be the sum T + C + S + I of the basic variables involved. Although we will be assuming the decomposition given by equation (I) as we examine the methods discussed in this chapter, analogous procedures are available when a sum of the degree of success achieved in applying the assumption.

Moving Averages; The Smoothing of Time Series

Given a set of numbers Y_1, Y_2, Y_3

We define moving average of order N to be the sequence of arthimetic means :

$$\frac{Y_1,Y_2,Y_3...Y_4}{N}, \frac{Y_2+Y_3+...Y_{N+1}}{N}, \frac{Y_3+Y_4+...Y_{N+2}}{N},...$$

The sums in the numerators of sequence are called moving totals of order N.

Problem No. 1: Given the numbers 2, 6, 1, 5, 3, 7, and 2, a moving average of order 3 is given by the sequence

$$\frac{2+6+1}{3}, \frac{6+1+5}{3}, \frac{1+5+3}{3}, \frac{5+3+7}{3}, \frac{3+7+2}{3} \quad \text{or } 3,4,3,5,4$$

It is customary to locate each number in the moving average at its appropriate position relative to the original data. In this example we would write

Original data	2,6,1,5,3,7,2
Moving average of order 3	3,4,3,5,4

each number in the moving average being the mean of the three numbers immediately above it.

If the data are given annually or monthly, a moving average of order N is called, respectively, an N-year moving average or an N-month moving average. Thus we speak of 5-year moving averages, 12-month moving averages, etc. Clearly, any other unit of time can also be used.

Moving averages have the property that they tend to reduce the amount of variation present in a set data. In the case of time series, this property is often used to eliminate unwanted fluctuations, and the process is called the smoothing of time series.

If weigted arthmetic menas are used in sequence (3), the weights being specified in advance, then the resulting sequence is call a weighted moving average of order N.

Problem No. 2 If the weights 1, 4, and 1 are used in Example 1, a weighted moving average of order 3 is given by the sequence.

$$\frac{1(2)+4(6)+1(1)}{1+4+1}, \frac{1(6)+4(1)+1(5)}{1+4+1}, \frac{1(1)+4+(5)+1(3)}{1+4+1},$$

$$\frac{1(5)+4(3)+1(7)}{1+4+1}, \frac{1(3)+4(7)+1(2)}{1+4+1} \quad \text{or } 4.5,\ 2.5,\ 4.0,\ 5.5.$$

Measure of Dispersion :

The word dispersion, literally, means deviation, difference or spread over of certain values from their central value. Further, the word 'measure' means a method of measuring or ascertaining certain values. Thus the phrase 'measures of dispersion' means the various possible methods of measuring the dispersions or deviations of the different values from a designated value of a series. The designated value may be an average value or any other extreme value. This measure indicates the degree or the extent to which there are differences between the various items and the designated value of a series.

The main objects of measuring dispersion are as :

i) To determine the reliability of an average.

ii) To serve as a basis for the control of variability.

iii) To compare two or more series with regard to their variability.

iv) To facilitate other statistical computation.

Measures of Dispersion

- Positional (Based on Limits)
  - Range
  - Quartile Deviation
- Arithmatic (Based on Difference)
  - Mean Deviation
  - Standard Deviation
- Graphic (Based on Graph)
  - Lorenz curve

I. Range

"Range represents the difference between the values of the extremes... the largest value and the smallest value. The values in between the two extremes are not at all taken into consideration.

The 'Range' gives an extremely simple indicator of the variability of a set of of observations. It is denoted symbolically by 'R'

(i) Range = largest value - smallest value

R = L-S ....Absolute Measure

(ii) Coefficient of Range = $\dfrac{\text{Largest value - Smallest value}}{\text{Largest value + Smallest value}}$

C. of R = $\dfrac{L - S}{L + S}$

Problem No. 1 : Compute the range and the Coefficient of Range of the series, and state which one is more dispersed and which one is more uniform.

Series	Values of variables
i	13, 14, 15, 16, 17 ... $(\overline{x} = 15)$ "Central tendency is same but formation differs"
ii	9, 12, 15, 18, 21... $(\overline{x} = 15)$
iii	1, 8, 15 22, 29.... $(\overline{x} = 15)$

Solution :

Range	(I)	(II)	(III)
	R = L-S	R = L-S	R = L-S
	= 7-3	= 21-9	= 29-1
	= 4	= 12	= 28

Coefficient of Range :

	cof oR	**cof oR**
$= \dfrac{L-S}{L+S}$	$= \dfrac{L-S}{L+S}$	$= \dfrac{L-S}{L+S}$
$= \dfrac{17-13}{17+13}$	$= \dfrac{21-9}{21+9}$	$= \dfrac{21-9}{21+9}$
$= \dfrac{4}{30}$	$= \dfrac{4}{30}$	$= \dfrac{28}{30}$
= 0.133	= 0.4	= 0.9333

Series (I) is less dispersed and more uniform..
Series (III) is **less** uniform **more** dispersed.

Problem No 2 : Compute the range and the Coefficient of Range of the series, and state which one is more dispered and which one is more uniform.

Series	Values of variables
A :	10, 11, 12, 13, 14... ($\overline{x}$ =12)
B :	40, 41, 42, 43, 44... ($\overline{x}$ =42)
C :	100,101, 102, 103, 104... ($\overline{x}$ =102)

"Central tendency is differs but but formation is same."

Solution :

Range

(I)
R = L–S
= 14–10
= 4

(II)
R = L–S
= 44–40
= 4

(III)
R = L–S
= 104–100
= 4

Coefficient of Range

(I)

$$\text{C of R} = \frac{L-S}{L+S} = \frac{14-10}{14+10} = \frac{4}{24} = 0.1667$$

(II)

$$\text{C of R} = \frac{L-S}{L+S} = \frac{44-40}{44+40} = \frac{4}{84} = 0.0476$$

(III)

$$\text{C of R} = \frac{L-S}{L+S} = \frac{104-100}{104+100} = \frac{4}{204} = 0.0196$$

Series (A) is **less** uniform and **more** dispersed.
Series (C) is **less** dispersed and **more** uniform.

Problem No. 3- From the following distribution find the range and the cofficient of Range.

X :	6	12	18	24	30	36	42
Y :	20	130	80	60	210	1500	600

Solution : In finding the range, the frequencies are never taken into account.

Range = L-S
= 42-6
= 36

$$\text{Cofficient of Range} = \frac{L-S}{L+S} = \frac{42-6}{42+6} = \frac{36}{48} = 0.75$$

Problem No. 4- Compute the range and the Coefficient of Range from the following distribution :

C.I :	120-130	130-140	140-150	150-160	160-170
F :	2	9	16	12	5

Solution : In finding the range, the frequencies are never taken into account. The upper limit of the highest class and the lower limit of the smallest class are only taken into account.

Range = L-S

= 170–120

= 50

Coefficient of Range = $\frac{L-S}{L+S} = \frac{170-120}{170+120} = \frac{50}{290} = 0.1724$

II. Quartile Deviation

It is nothing but a "Semi-interquartile Range". With some modifications, it is similar to the range. In the distribution, we consider Q_3 as the largest value and Q_1 as the smallest. It means, the items below the lower quartile and the items above the upper quartile are not at all included in the computation. Thus we are considering only the middle half protion of the distribution. The range so obtained is divided by two as we are considering only half of the data. Thus the Quartile deviation measures the difference between the values of Q_1 and Q_3. It is denoted symbolically by 'Q.D.'

a) Quatile Deviation

$$Q.D = \frac{Q_3 - Q_1}{2} \quad \text{...Absolute Measure.}$$

b) Coefficient of C of Q.D $= \frac{Q_3 - Q_1}{Q_3 - Q_1}$ ...Relative Measure.

For example, if $Q_1 = 15$ and $Q_3 = 40$

$$Q.D. = \frac{Q_3 - Q_1}{2}$$

$$= \frac{40-15}{2}$$

$$= \frac{25}{2} = 12.5$$

$$C \text{ of } Q.D. = \frac{40-15}{40+15}$$

$$= \frac{25}{55} = 45.45$$

The Quartile Deviation gives the average amount by which the two quartiles differ from median in an asymmetrical distribution. Rigorously speaking quartile deviation is only a positional average. It does not exhibit any scatter around an average. It is a measure of partition rather than a measure of dispersion. The smallest value of Q.D., the minimum is the dispersion of middle half of the distribution around the median. However, it provides on indication of the, degree of dispersion lying beyond the limits of the two quartiles.

Problem No. 1 From the following marks of the 12 students, compute the quartile Deviation and its Coefficient.

S.No. :	1	2	3	4	5	6	7	8	9	10	11	12
Marks:	25	30	37	43	48	54	61	67	72	80	84	89

Solution : **Compution of Quartation of Quartile Deviation**

S.No.	Marks
1	25
2	30
3	37
4	43
5	48
6	54
7	61
8	67
9	72
10	80
11	84
12	89

Q_1 is the size of $= \frac{n+1}{4}$ the item.

$= \frac{12+1}{4}$

$= 3.25$th item.

$Q_1 = 3\text{rd item} + 0.25\ (43-37)$

$= 37+0.25\ (43-37)$

$= 37 + 0.25\ (6)$

$= 37 + 1.5$

$= 38.5$

Q_3 is the size of th $= \frac{3(n+1)}{4}$ th item.

$= \frac{3(12+1)}{4}$

$= 9.75$th item.

$Q_3 = 9\text{th item} + 0.75\ (80-72)$

$= 72+0.75\ (80-72)$

$= 72+0.75\ (8)$

$= 72 + 6$

$= 78$

$$\text{Q.D.} = \frac{Q_3 - Q_1}{2}$$

$$= \frac{78 - 38.5}{2}$$

$$= \frac{39.5}{2} = 19.75$$

$$\text{Co-efficient of Q.D.} = \frac{Q_3 - Q_1}{Q_3 - Q_1}$$

$$= \frac{78 - 38.5}{78 + 38.5}$$

$$= \frac{39.5}{116.5} = 0.339$$

Problem No. 2- Calculate the Quartile Deviation and its Coefficient from the following data :

X	:	58	59	60	61	62	63	64	65	66
F	:	15	20	32	35	33	22	20	10	8

Solution : Compution of Quartation of Quartile Deviation

	X	F	C.F
	58	15	15
	59	20	35
(Q_1)	60	32	67
	61	35	102
	62	33	135
(Q_3)	63	22	157
	64	20	177
	65	10	187
	66	8	195
	n = 195		

Q_1 is the size of $= \frac{n+1}{4}$ the item.

$$= \frac{195+1}{4} = 48.78\text{th.}$$

It lies in 67 c.f. Against 67 C.f.. $Q_1 = 60$

Q_3 is the size of th $= \frac{3\ (n+1)}{4}$ th item.

$$= \frac{3\ (195+1)}{4}$$

$$= 146.33\text{th.}$$

It lies 157 c.f. Against 157 c.f.. $Q_3 = 63$

$$\text{Q.D.} = \frac{Q_3 - Q_1}{2} = \frac{63-60}{2} = \frac{3}{2} = 1.5$$

$$\text{Coefficient of Q.D.} = \frac{Q_3 - Q_1}{Q_3 - Q_1} = \frac{63-60}{63+60} = \frac{3}{123} = 0.02439$$

Problem No. 3 - From the following compute Quartile Deviation and the Coefficient of Quartile Deviation.

Weekly Wages (Rs.)	: 4-8	8-12	12-16	16-20	20-24	24-28	28-32	32-36	36-40
No. of Workers:	6	10	18	30	15	12	10	6	2

Solution : Compution of Quartile Deviation

	Weekly Wages (Rs.)	No. of Workers (f)	C.f.
	4-8	6	6
	8-12	10	16 (m)
(l)	12-16	18(f)	34 ← Q_1
	16-20	30	64
	20-24	15	79 (m)
(l)	24-28	12(f)	91 ← Q_3
	28-32	10	101
	32-36	6	107
	36-40	2	109
		n=109	

Q_1 is the size of $\frac{n}{4}$ th them $\therefore \frac{109}{4}$ th = 27.25th item.

It lies in 34 c.f. Against 34 c.f. the Q_1 class is (12-16)

Q_3 is the size of $\frac{3n}{4}$ th item. $\therefore \frac{3(109)}{4}$ th = 81.75th items

It lies in 91 c.f. Against 91 c.f. the Q_3 class is (24-28).

$$Q_1 = L + \frac{\frac{n}{4} - m}{f} \times c = 12 \; \frac{\frac{109}{4} - 16}{18} \times 4 = 12 + \frac{27.25 - 16}{18} \times 4$$

$$= 12 + \frac{11.25}{18} \times 4 = 12 + \frac{45}{18} = 12 + 2.5 = 14.5$$

$$Q_3 = L + \frac{\frac{3n}{4} - m}{f} \times c = 12 \; \frac{\frac{3(109)}{4} - 79}{12} \times 4 = 24 + \frac{81.75 - 79}{12} \times 4$$

$$= 24 + \frac{2.75}{12} \times 4 = 24 + \frac{11}{12} = 24 + 0.9167 = 24.9167$$

$$QD = \frac{Q_3 - Q_1}{2} = \frac{24.9167 - 14.5}{2} = \frac{10.4167}{2} = 5.208$$

Coefficient of Q.D. =

$$= \frac{Q3 - Q1}{Q3 - Q1} = \frac{24.9167 - 14.5}{24.9167 + 14.5} = \frac{10.4167}{39.4167} = 0.26427$$

Mean-Deviation or Average-Deviation or Mean Absolute Deviation (δ =Delta)

The range and quartile deviation are not based on all observations. They are positional measures of dispersion.

M.D. is the arthmetic mean of the deviation of a series computed from any measure of central tendency i.e. the mean, median or mode, all the deviation are taken as positive i.e. (+) and (-) sign are ignored.

Mean-devations is denoted by the Greek letter. Small δ (Delta) we usally compute. M.D. about any one of the three averages mean, median or mode. Sometimes mode may be ill-defined and as such mean-deviation is computed from mean and median. Median is preferred as a choice between mean and median. But in general practice and due to wide applications of mean, the mean deviation is generally computed from mean.

Calculation of Mean-Deviation and It's Coefficient in Simple Series

The following steps are taken for calculating M.D. in Individual Series.

(1) Calculate the Mean, Median and Mode of the Series.

(2) Take the deviations of the items from the Mean, Median and Mode ignoring $\pm$ sign and denote them by $d\overline{X}$, dm and dz.

(3) Obtain the total of these deviations i.e. $\sum d\overline{X}$, $\sum dm$, $\sum dz$

(4) The formula is as -

(a) M.D. from Mean $(\delta\overline{X}) = \dfrac{\sum \delta\overline{X}}{n}$

$$\text{Coefficient of } (\delta\overline{X}) = \frac{\delta\overline{X}}{X}$$

Where, $\delta\overline{X}$ = Mean-Deviation from mean and $\overline{X}$/a =Mean.

(b) M.D. from Median $(\delta m) = \dfrac{\sum dm}{n}$

$$\text{Coefficient of M.D. from Median} = \frac{\delta m}{m}$$

Where, δ m = M.D. from Median, m=Median

(c) M.D. from Mode(δ z) $= \dfrac{\sum dz}{n}$

$$\text{Coefficient of M.D. from Median} = \frac{\delta z}{z}$$

Where, δ z = M.D. from Mode, z = Mode

Prooblem 1 : Calculate Mean-Deviation from mean and it's coefficient.

Marks : 20 22 27 30 31 32 35 40 45 48

Marks	dx=30	$d\overline{X}$ =33 ($\pm$ Ignored)
20	–10	13
22	–8	11
27	–3	6
30	0	3
31	1	2
32	2	1
35	5	2
40	10	7
45	15	12
48	18	15
N=10	$\sum dx = 30$	$\sum d\overline{X} = 72$

$$\text{Mean } (\bar{X}) = x + \frac{\Sigma dx}{n}$$

$$(\bar{X}) = 30 + \frac{30}{10}$$

$$(\bar{X}) = 30 + 3$$

$$(\bar{X}) = 33$$

$$\text{M.D. from mean} (\delta \overline{X}) = \frac{\Sigma d\overline{X}}{n} = \frac{72}{10}$$

$$(\delta \overline{X}) = 7.2$$

$$\text{Coefficient of M.D. from mean} = \frac{\delta \overline{X}}{X}$$

$$= \frac{7.2}{33} = 0.218$$

Coefficient of M.D. from mean = 0.218

Problem 2 : Calculate Mean-Deviation from Median and its's coefficients.
Income in Rs. : 10 12 15 20 25 30 35 42 40

Marks	Ascending order	dm=25 ($\pm$ Ignored)
10	10	15
12	12	13
15	15	10
20	20	5
25	25	0
30	30	5
35	35	10
42	40	15
40	42	17
	N=9	$\Sigma dm = 90$

$$\text{Mdian } (M) = \text{The size of} \left(\frac{n+1}{2}\right)^{th} \text{Item}$$

$$(M) = \text{The size of} \left(\frac{9+1}{2}\right)^{th} \text{Item}$$

$$(M) = \text{The size of} \left(\frac{10}{2}\right)^{th} \text{Item}$$

(M) = The size of 5th Item in A.O. is 25
M = 25

$$\text{M.D. from Median} = (\delta m) = \frac{\sum dm}{n}$$

$$\delta m = \frac{90}{9} = 10$$

$$\text{Coefficient of M.D. from Median} = \frac{\delta m}{m} = \frac{10}{25} = 0.4$$

Problem 3 : Calculate Mean-Devation from Mode and it's coefficients.

Marks : 20 14 8 20 18 30 24 14 18 14

Solution :

Marks	dz = 14 ($\pm$ Ignored)
14	0
8	6
20	6
18	4
30	16
24	10
14	0
18	4
14	0
N = 4	$\sum dz = 46$

Mode (Z) = By observation
14 is repeated 3 times

M.D. from Mode -

$$(\delta z) = \frac{\sum dz}{n}$$

$$(\delta z) = \frac{46}{9}$$

$$\delta z = 5.111$$

$$\text{Coefficient of M.D. from Mode} = \frac{\delta z}{z} = \frac{5.111}{14} = 0.37$$

IV. Standard Deviation (σ)

'Standard deviation' is the root of the sum of the squares of the deviations devided by their number. It is also called "Mean Error Deviation", 'Mean Square Error Deviation' or "Root Mean Square Deviation." It is a second moment of a dispersion. Since the sum of the squares of the deviations from the mean is minimum, the deviations are taken only from mean but not from median or mode).

Standard Deviation is the root-mean-square average of all the deviations from the mean. It is proposed by **Prof. Karl Pearson** in 1893, and it is denoted by 'σ' (sigma)

A: Individual Observations :

$$\sigma = \sqrt{\frac{\sum d^2}{n}} \text{ or } \sqrt{\frac{\sum (x-\bar{x})^2}{n}} \quad \text{.....Absolute Measure}$$

$$\text{Coefficient of Variation } \frac{\sigma}{x} \times 100 \quad \text{......Relative Measure}$$

B: Discrete and Continuous Series :

$$\sigma = \sqrt{\frac{\sum fd^2}{n}} \text{ or } \frac{\sum f \ (x-\bar{x})^2}{n} \quad \text{.....Absolute Measure}$$

Coefficient of Variation = $\frac{\sigma}{n} \times 100$ .....Relative Measure

Step to be followed :

(i) Calculate the $\overline{x}$

(ii) Note down the deviations $\left[\text{i.e.,} (x - \overline{x}) \text{or d}\right]$

(iii) Square the deviations $\left[\text{i.e.,} (x - \overline{x})^2 \text{ or } d^2\right]$

(iv) Multiply d^2 by f (i.e., fd^2)

(v) Total the fd^2 $\left(i.e., \sum fd^2\right)$

(vi) Divide $\sum fd^2$ by n $\left(\text{i.e.,} \sum fd^2/n\right)$

(vii) Obtain the roots of $\sum fd^2/n$ $\left(\text{i.e.,} \sqrt{\sum fd^2/n}\right)$

Short-cut Method

Simetimes the mean will be a fractional figure. Then we should take the deviations from the assumed mean and the direct method formula will be having some adjustment. As the deviations are nor taken from the actual mean, we get the '$\sum d$' as some value instead zero. The shor cut formula works as under,

A : Individual observations : $\sigma = \sqrt{\frac{\sum d^2}{n} - \sqrt{\left(\frac{\sum d}{n}\right)^2}}$

B : Discrete and Continues series : $\sigma = \sqrt{\frac{\sum d^2}{n} - \sqrt{\left(\frac{\sum d}{n}\right)^2}}$

Step-deviation Method

The deviations are further divided by the common factor in case of assumed mean. This deliberate error is compensated by multiplying the entire formula by the same factor. The formula works as under,

A : Individual Obsevations : C = Common Factor

$$\sigma = \sqrt{\frac{\sum d'^2}{n} - \left(\frac{\sum d'}{n}\right)^2} \times c$$

B : Discrete and Continues series :

$$\sigma = \sqrt{\frac{\sum d'^2}{n} - \left(\frac{\sum d'}{n}\right)^2} \times c$$

Problem 1 : Following are the marks obtained by two students 'A' and 'B' in 10 tests of 100 marks each.

Test :	1	2	3	4	5	6	7	8	9	10
A :	44	80	76	48	52	72	72	51	60	54
B :	48	75	54	60	63	69	72	51	57	66

Find who is better in studies and if consistency is the criterion for aswarding a prize, who should get the prize.

Solution :

Compution of Mean and Standard Deviation

A	$(x-\bar{x})$	d²	B	$(x-\bar{x})$	d²
44	–16.9	285.61	48	–13.5	182.25
48	–12.9	166.41	51	–10.5	110.25
51	–9.9	98.01	54	–7.5	56.25
52	–8.9	79.21	57	–4.5	20.25
54	–6.9	47.61	60	–1.5	2.25
60	–0.9	0.81	63	+1.5	2.25
72	+11.1	123.21	66	+4.5	20.25
72	+11.1	123.21	69	+7.5	56.25
76	+15.1	228.01	72	+10.5	110.25
80	+19.1	364.81	75	+13.5	182.25
$\sum x =$ 609	$\sum d = 0$	$\sum d^2 =$ 1516.90	$\sum x =$ 615	$\sum d = 0$	$\sum d^2 =$ 742.50

$$\bar{x} = \frac{\sum x}{n} : \sigma = \sqrt{\frac{\sum d^2}{n}} : \text{c.v.} = \frac{\sigma}{\bar{x}} \times 100$$

$$= \frac{609}{10} : = \sqrt{\frac{1516.90}{10}} : = \frac{12.316}{60.9} \times 100$$

$$= 60.9 : = \sqrt{151.69} : = 20.22\%$$

$$= 12.316$$

$$\bar{x} = \frac{\sum x}{n} : \sigma = \sqrt{\frac{\sum d^2}{n}} : \text{c.v.} = \frac{\sigma}{\bar{x}} \times 100$$

$$= \frac{615}{10} : = \sqrt{\frac{742.50}{10}} : = \frac{8.617}{61.5} \times 100$$

$$= 61.5 : = \sqrt{74.25} : = 14.01\%$$

$$= 8.617$$

B is better in studies and he should get a prize also as his average is more and variance is less.

Problem 2 : Following are the runs scored by two batsmen A and B :

A :	70	90	80	50	40
B :	70	90	60	50	30

Find who is a better run getter and more consistent plyer.

Solution : **Compution of Quartation of Quartile Deviation**

A	$(x-\bar{x})$	d^2	B	$(x-\bar{x})$	d^2
X		d	X	d	
70	+4	16	70	+10	100
90	+24	576	90	+30	900
80	+14	196	60	0	0
50	–16	256	50	–10	100
40	–26	676	30	–30	900
$\Sigma d = 330$	$\Sigma d = 0$	$\Sigma d^2 = 1720$	$\Sigma d = 300$	$\Sigma d = 0$	$\Sigma d^2 = 2000$

$$\bar{x} = \frac{\Sigma x}{n}\ \sigma = \frac{\Sigma d^2}{n}\ \text{c.v.} = \frac{\sigma}{\bar{x}} \times 100 \qquad \bar{x} = \frac{\Sigma x}{n} : \sigma = \sqrt{\frac{\Sigma d^2}{n}} : \text{c.v.} = \frac{\sigma}{\bar{x}} \times 100$$

$$= \frac{330}{5} : = \sqrt{\frac{1720}{5}} : = \frac{18.547}{66} \times 100 \qquad = \frac{300}{5} = \sqrt{\frac{2000}{5}} = \frac{20}{60} \times 100$$

$$= 66 : = \sqrt{344} : = 28.10\% \qquad = 60 = \sqrt{400} = 33.33\%$$

$$= 18.547 \qquad = 20$$

Batsman A is better run getter and more consistent as his average is more and variation is less.

Pronlem No. 3 :

From the following data find σ by the direct method proper:

X: 8 9 15 23 5 11 19 8 10 12

Solution :

Computation of the Standard Deviation by the direct method proper

	X	$(X-\bar{X})$=**x**	**x^2**
1	8	-4	16
2	9	-3	9
3	15	3	9
4	23	11	121
5	5	-7	49
6	11	-1	1
8	7	19	49
8	-4	16	16
9	10	-2	4
10	12	0	0
N=10	120	-	$\Sigma x^2 = 274$

Where, $\bar{X} = \frac{\Sigma X}{N} = \frac{120}{10} = 12$

$$\sigma = \frac{\Sigma x^2}{N} = \sqrt{\frac{274}{10}} = \sqrt{27.4} = 5.23$$

Ptoblem 4 :

Find the standard deviation from the following group data by the direct method proper.

X:	5-1.5	1.5-2.5	2.5-35	3.5-4.5	4.5-5.5	5.5-6.5	6.5-7.5
F:	7	6	5	4	3	2	1

Solution

Computation of the Standard Deviation by the direct method proper

Class Marks	F	m	Fm	$(m-\bar{x})$	x^2	Fx^2
.5-1.5	7	1	7	-2	4	28
1.5-2.5	6	2	12	-1	1	68
2.5-3.5	5	3	15	0	0	08
3.5-4.5	4	4	16	1	1	48
4.5-5.5	3	5	15	2	4	128
5.5-6.5	2	6	12	3	9	188
6.5-7.5	1	7	7	4	16	168
Total	N=28		84	-	-	848

$$\bar{x} = \frac{\Sigma Fm}{N} = \frac{84}{28} = 3$$

$$\sigma = \sqrt{\frac{\Sigma Fx^2}{N}}$$

$$= \sqrt{\frac{848}{28}}$$

$$= \sqrt{30.28} = 5.50 \text{ αpprox.}$$

Problem 5: The following table gives the age distribution of boys and girls in a high school. Find which of the two groups is more variable in age.

Age in years :	13	14	15	16	17
No. of students					
(boys):	12	15	15	5	3
(girls) :	13	10	12	2	1

Solution : **Computation of Standard Deviation**

Age (x) in years	f	fx	Boys		
			d	fd	fd^2
13	12	156	-2	24	48
14	15	210	-1	-15	15
15	15	225	0	0	0
16	5	80	+1	+5	5
17	3	51	+2	+6	12
n=50	$\sum fx=722$			$\sum fd=-28$	$\sum fd^2=-80$

$$\bar{x} = \frac{\sum fx}{n}$$

$$= \frac{722}{50} = 14.44$$

$$\sigma = \sqrt{\frac{\sum fd^2}{n} - \left(\frac{\sum fd}{n}\right)^2}$$

$$= \sqrt{\frac{80}{50} - \left(\frac{-28}{50}\right)^2}$$

$$= \sqrt{1.6 - (-0.56)^2}$$

$$= \sqrt{1.6 - 0.3136}$$

$$= \sqrt{1.28641} = 1.1342$$

$$\text{c.v.} = \frac{\sigma}{\bar{x}} \times 100 = \frac{1.342}{14.44} \times 100 = 7.855\%$$

Computation of Standard Deviation

Age (x) in years	f	fx		Girls	
			d	fd	fd^2
13	13	169	-2	-26	52
14	10	140	-1	-10	10
15	12	180	0	0	0
16	2	32	+1	+2	2
17	1	17	+2	+2	4
	n=38	$\sum fx=538$		$\sum fd=-32$	$\sum fd^2=-68$

$$\bar{x} = \frac{\sum fx}{n} = \frac{538}{38} = 14.16$$

$$\sigma = \sqrt{\frac{\sum fd^2}{n} - \left(\frac{\sum fd}{n}\right)^2}$$

$$= \frac{68}{38} - \left(\frac{-32}{38}\right)^2$$

$$= \sqrt{1.7895 - (-0.842)^2}$$

$$= \sqrt{1.7895 - 0.7091}$$

$$= \sqrt{1.0804} \quad = 1.039$$

$$\text{c.v.} = \frac{\sigma}{\bar{x}} \times 100 = \frac{1.039}{14.16} \times 100 = 7.341\%$$

Age of boys is more variable as its variable as its variation is more.

Problem 6 : A factory produces two types of tyres. In an experiment in the following in the working life of these tyres the following results are obtained :

Length of life : in '00' hours.	15-17	17-19	19-21	21-23	23-25
Type A :	50	110	260	100	80
Type B :	40	300	120	80	60

State which type of tyres are more stable.

Solution : Computation of Standard Deviation

Lenght of life in '00' hours fd² C.I.	x	(x-20)/2 d'	A type			B type		
				f	fd'	fd'²	f	fd'
15-17	16	-2	50	-100	200	40	-80	160
17-19	18	-1	110	-110	110	300	-300	300
19-22	20	0	260	0	0	120	0	0
21-23	22	+1	100	+100	100	80	+80	80
23-25	24	+2	80	+160	320	60	+120	240
		$\sum d = 0$	$n = 600$	$\sum fd' =$ 50	$\sum fd'2 =$ 730	$n =$ 600	$\sum fd'2 =$ 180	$\sum fd'2 =$ 780

$$\bar{x}_A = A + \frac{\sum fd'}{n} \times c = 20 + \frac{50}{600} \times 2 = 20 + 0.1667 = 20.1667$$

$$\bar{X}_B = A + \frac{\Sigma fd'}{n} \times c = 20 + \frac{-180}{600} \times 2 = 20 - 0.6 = 19.4$$

$$\sigma_A = \sqrt{\frac{\Sigma fd'2}{n} - \left(\frac{\Sigma fd'}{n}\right)^2} \times c = \sqrt{\frac{730}{600} - \left(\frac{50}{600}\right)^2} \times 2$$

$$\sqrt{1.20976} \times 2 = 1.099891 \times 2 = 2.1998$$

$$\text{c.v. } \frac{\sigma}{\bar{x}} \times 100 = \frac{2.1998}{20.1667} \times 100 = 10.91\%$$

$$_B = \sqrt{\frac{\Sigma fd'2}{n} - \left(\frac{\Sigma fd'}{n}\right)^2} \times c = \sqrt{\frac{780}{600} - \left(\frac{-180}{600}\right)^2} \times 2$$

$$= \sqrt{1.3 - (-0.3)^2} \times 2 = \sqrt{1.3 - 0.09} \times 2 = \sqrt{1.21} \times 1.1 \times 2 = 2.2$$

$$\text{c.v.} = \frac{\sigma}{\bar{x}} \times 100 = \frac{2.2}{19.4} \times 100 = 11.34\%$$

Type **'A'** tyers are more stable as their variation is less than **'B'** tyres.

Problem 7: An agent obtained samples of bulbs from the 2 companies. He had them tested for durability and got the following results :

Durability in '00' hr.	Company A	Company B
17-19	100	30
19-21	160	420
21-23	260	120
23-25	80	30

Which company bulbs are more uniform?

Solution : **Computation of Standard Deviation**

Lenght of life in '00' hours C.I.	x	(x-20)/2 d'	A Company f	fd'	fd'2	B Company f	fd'	fd2
17-19	18	-1	100	-100	200	30	-30	30
19-21	20	0	160	0	110	420	-0	0
21-23	22	+1	260	+260	260	120	+120	120
23-25	24	+2	80	+160	320	80	+60	120
			n = 600	$\Sigma fd'$ = 320	$\Sigma fd'2$ = 680	n = 600	Σ fd'2 = 150	Σ fd'2 = 270

$$\bar{x}_A = A + \frac{\sum fd'}{n} \times c$$

$$= 20 + \frac{320}{600} \times 2$$

$$= 20 + 1.0667$$

$$= 21.0667$$

$$x_B = A + \frac{\sum fd'}{n} \times c = 20 + \frac{150}{600} \times 2$$

$$= 20 + 0.5 \qquad = 20.5$$

$$\sigma_A = \sqrt{\frac{\sum fd'^2}{n} - \left(\frac{\sum fd'}{n}\right)^2} \times 2$$

$$= \frac{680}{600} - \left(\frac{320}{600}\right)^2 \times 2$$

$$= \sqrt{1.1333 - (0.533)^2} \times 2$$

$$= \sqrt{1.1333 - 0.2844} \times 2$$

$$= \sqrt{0.84885} \times 2$$

$$= 0.92136 \times 2$$

$$= 1.8427$$

$$\text{c.v.} = \frac{\sigma}{\bar{x}} \times 100$$

$$= \frac{184.27}{2106.67} \times 100$$

$$= 8.747\%$$

$$\sigma_B = \sqrt{\frac{\sum fd'2}{n} - \left(\frac{\sum fd'}{n}\right)^2} \times c$$

$$= \sqrt{\frac{270}{600} - \left(\frac{150}{600}\right)^2} \times 2$$

$$= \sqrt{0.45 - (-0.25)^2} \times 2$$

$$= \sqrt{0.45 - 0.0625} \times 2$$

$$= \sqrt{0.3875} \times 2$$
$$= 0.622495 \times 2$$
$$= 1.245$$

$$\text{c.v.} = \frac{\sigma}{\overline{x}} \times 100 = \frac{1.245}{20.5} \times 100 = 6.073\%$$

The Bulbs of **'B'** company are more uniform and durable than the bulbs of **'A'** company as the variation in B bulbs is less.

Lorenz Curve

This curve was used for the first time for measring the distribution of Wealth and Income. The following steps are to be taken up in turn for drawing such curves :

1. Cumulate both values and their corresponding frequencies.
2. Find the percentages of the cumulated figures taking the grand total of each corresponding column as 100.
3. Represent the percentages of each ofthe cumulative frequencies on X-axis and those of the values on the Y-axis.
4. Draw a diagonal line designated as the "Line of equal distribution" joining the point O (0,0) with the point p (100 100) at an angle of 45^0 on the Y-axis.
5. Plot the percentages of the cumulated values against the percentages of the cumulated frequencies of a given distribution and join the points so plotted through a free-hand curve. The greater the distance of this curve from the "line of equal distribution" the greater is the variability of the distribution.

A specimen of the Lorenz curve described above is presented here as under :

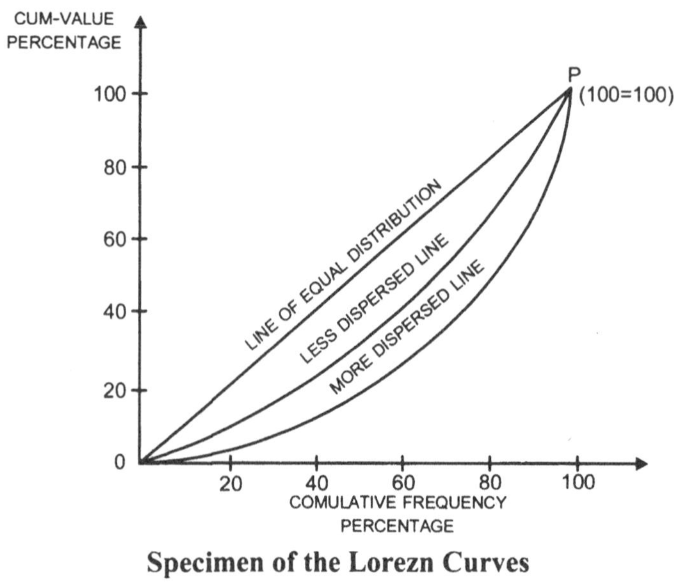

Specimen of the Lorezn Curves

Line of equal distribution

The line of equal distribution is a diagonal line on the Lorenz curve on which any point shows the same per cent on both x and y axis.

Problem 1 : From the following data relating to the purchases made by the customers of the two different localities study the dispersions by means of the Lorenz curves.

Purchase in '000 Rs.	6	25	60	84	105	150	170	400
No. of Customers :								
Locality X :	6	11	13	14	15	17	10	14
Locality Y :	2	38	52	28	38	26	12	4

Solution : Computation of the Lorenz factors

	Purchases			**Locality X**			**Locality Y**		
	Amount in '000 Rs. X	Cum X	**Cum** % of X	**No** customers F	CF	CF%	**No. of** custom ers F	CF	CF%
	6	6	0.6	6	6	6	2	2	1
	25	31	3.1	11	17	17	38	40	20
	60	91	9.1	13	30	30	52	92	46
	84	175	17.5	14	44	44	28	120	60
	105	280	28.0	15	59	59	38	158	79
	150	430	43.0	17	76	76	26	184	92
	170	600	60.0	10	86	86	12	196	98
	400	1000	100	14	100	100	4	200	100
Total	1000	=	100	100	=	100	200	=	100

Lorenz Curves

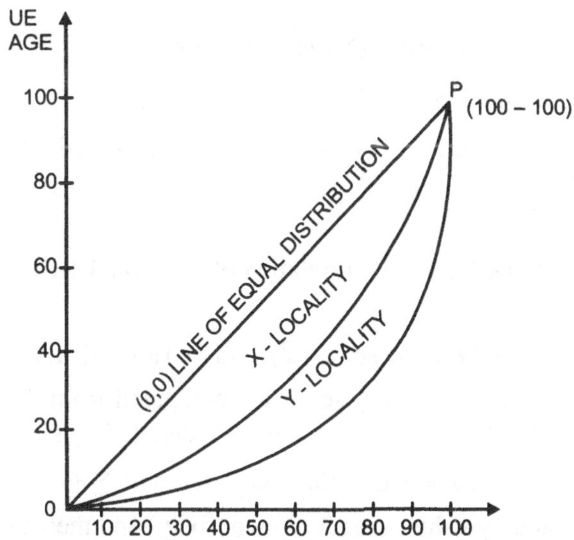

Cumulative Frequency Percentage

Since the curve of the locality Y is farthest fom the line of the equal distribution, it represents greater dispersion or inconistency.

Correlation Analysis :

Correlation analysis is defined as the statistical technique which measures the degree of relationship (or association) between variables. In other words, it helps us in stydying the covariance of two or more variables.

According to W.I. Kings, "Correlation means that between two series or groups of data there exists some casual connections."

According to A. M. Tuttle, "Correlation is an analysis of the covariation between two or more variables."

(1) Kark Pearson's Coefficient of Correlation

Karl Pearson (1857-1936) was a great biometrician and statistician. Suggested a Mathematical method for measuring the magnitude of Linear relationship between two variables. Karl Pearson's method is the most widely used method in practice and is known as Pearsonian coefficient of correlation or covariance method. It is denoted by the symbol 'r' (small r). The formula is-

$$r = \frac{\text{Covariance of } xy}{SD_X \times SD_y} \quad \text{... (1)}$$

$$r = \frac{\Sigma xY}{n \times SD_X \times SD_y} \quad \text{... (2)}$$

$$r = \frac{\Sigma xY}{\Sigma x^2 \times \Sigma y^2} \quad \text{... (3)}$$

$$x = (x - \bar{X})$$

$$Y = (y - \bar{Y})$$

$$(\sigma)\ SDx = Standard - Deviation\ of\ x\ series$$

$$(\sigma)\ SD_y = Standard - Deviation\ of\ y\ series$$

Karl Pearson coefficient of correlation can be computed either by Direct method or Short-cut Method.

Direct Method

To calulate K.P. coefficient of correlation of individual series the following process is involved.

(i) Arithmetic Mean of both the series $(\bar{x})$ and $(\bar{y})$ are calculated.

(ii) Deviation of the x series and y series are computed from their respective arithmetic mean for all individual values. Deviation are denoted by symbols dx and dy.

(iii) Deviations are squared added to find out Σdx^2 and Σdy^2.

(iv) Individual dx and dy values are multiplied (dxdy) and their total (Σdxdy) is obtained.

(v) Standard-Deviation (S.D.) of both the series are computed from the following formula-

$$SD_x \text{ or } \delta\ x = \sqrt{\frac{\Sigma dx^2}{n}} \qquad SD_y \text{ or } \delta\ y = \sqrt{\frac{\Sigma dy^2}{n}}$$

(vi) Coefficiemt of correlatopm is obtained from the following formula-

$$r = \frac{\Sigma dxdy}{n \times \delta \ x \times \ \delta \ y}$$

(vii) This direct method can further be simplified if instead of calculating S.D. of the series, the various values are taken directly to the formula for coefficient of correlation. Formula of calculating 'r' can thus be written as

$$r = \frac{\Sigma dxdy}{\sqrt{\frac{\Sigma dx^2}{n} \times \frac{\Sigma dy^2}{n}}} \text{ or } r = \frac{\Sigma dxdy}{\sqrt{\Sigma dx^2 \times \Sigma dy^2}}$$

Where $\Sigma dxdy$ = Total of the products of the deviation of values from the repective arthmetic mean in both the series.

N = Number of items

δx = S.D. of x series

δy = S.D. of y series

Problem 1 : Calculate Karl Pearson's coefficient of correlation by Direct Method

Capital 10 20 30 40 50 60 70 80 90 100

Profit 2 4 8 5 10 15 14 20 22 30

Solution : It is assumed that Capital = x series and Profit = y series

x	dx=55	dx^2	y	dy=13	dy^2	dxdy
10	–45	2025	2	–11	121	495
20	–35	1225	4	–9	81	315
30	–25	625	8	–5	25	125
40	–15	225	5	–8	64	120
50	–5	25	10	–3	9	15
60	5	25	15	2	4	10
70	15	225	14	1	1	15
80	25	625	20	7	49	175
90	35	1225	22	9	81	315
100	45	2025	30	17	289	765
N=10		Σdx^2=8250	N=10		Σdy^2=724	$\Sigma dxdy$=2350
$\Sigma x = 550$			$\Sigma y = 130$			

Mean x series

$$\overline{X} = \frac{\Sigma x}{n}$$

$$\overline{X} = \frac{550}{10} = 55$$

S.D. of x series

$$\delta x = \sqrt{\frac{\Sigma dy^2}{n}}$$

$$\delta x = \sqrt{\frac{8250}{10}}$$

$$\delta x = \sqrt{825}$$

$$r = \delta x = 28.722$$

Mean y series

$$\overline{X} = \frac{\Sigma y}{n}$$

$$\overline{X} = \frac{130}{10} = 13$$

S.D. of x series

$$\delta x = \sqrt{\frac{\Sigma dy^2}{n}}$$

$$\delta x = \sqrt{\frac{724}{10}}$$

$$\delta x = \sqrt{72.4}$$

$$\delta x = 8.509$$

Then by applying K.P. coefficient of correlation

$$r = \frac{\Sigma\, dxdy}{n \times \delta x \times \delta y}$$

$$r = \frac{2350}{10 \times 28.722 \times 8.509}$$

$$r = \frac{2350}{2443.95}$$

$$r = 0.96$$

(OR)

$$r = \frac{\Sigma\, dxdy}{\sqrt{\Sigma dx^2} \times \sqrt{\Sigma dx^2}}$$

$$r = \frac{2350}{\sqrt{8250 \times 724}}$$

$$r = \frac{2350}{\sqrt{59.73000}}$$

$$r = \frac{2350}{2443.97} \qquad r = 0.96$$

Short-Cut Method

In calcuting coefficient of correlation by short-cut method, the following steps are taken :

(i) Some convenient value are taken as assumed mean in series x and y.

(ii) Deviations are found out from the assumed mean of the x series and y series from their respective individual values. Deviation are denoted by symbols dx and dy.

(iii) Deviations are added up to find Σdx and Σdy

(iv) Individual dx and dy values are multiplied and their total $(\Sigma dxdy)$ is obtained.

(v) Deviations (dx and dy) an squared up and their total Σdx^2 and Σdy^2 are obtained.

(iv) Apply the following formula.

$$r = \frac{n \times \Sigma dxdy - (\Sigma dx \times \Sigma dy)}{\sqrt{n \times \Sigma dx^2 - (\Sigma dx)^2} \times \sqrt{n \times \Sigma dy^2 - (\Sigma dy)^2}}$$

When,

N = Number of items

$\Sigma dxdy$ = Total of product of the deviations of value from the respective arithmetic mean in both the series.

Σdx, Σdy = Sum of the deviation from the assumed mean in x and y series respectively.

$\Sigma dx^2 = \Sigma dy^2$ Sum of the square of the deviations from the assumed mean in x and y series respectively.

Formula for Log

$$r = \text{Antilog of } \left[\log A - \frac{\log B + \log c}{2}\right]$$

Probable Error (P.E.)

Probable Error of the correlatin coefficient applicable for the measurement of reliability of the computed value of the correlation coefficient. the probable Error (P.E.) is defined by

$$\text{P.E.} = 0.6745 \times \frac{1-(r)^2}{\sqrt{n}}$$

When,

0.6745 is a constant number

r = Pearsonian coefficient of correlation

N = No. of pairs of observations.

Again,

$$\frac{1-(r)^2}{\sqrt{n}} = \text{S.E.} = \text{Standard Error of correlation coefficient.}$$

Rules of Significant of Coefficient of Correlation

It is compared with 6 times of Probable Error (P.E.)

(a) If coefficient of correlation (r) is more than or greater than 6 times of P.E. the coefficient of correlation (r) is significant or $r > 6.\text{P.E.}$

(b) If coefficient of correlation (r) is less than or smaller than 6 times of P.E. the coefficient of correlation (r) is unsignificant or $r < 6.\text{P.E.}$

Probable Limits

The upper and lower limits within which the correlation coefficient in the population is expected to lie are (r+P.E.) and (r-P.E.) respectively.

Thus,

Maximim limit = r + P.E.

Minimum limit = r - P.E.

Properties of Correlation Ceofficient

The following are the important properties of correlation coefficients

(1) The values of r has no units.

(2) Its formula is symmetrical i.e. $r_{xy} = r_{yx}$

(3) Its value lie between - 1 to 1 i.e. $-1 \leq r \leq 1$

(4) If r=0 there is no correlation.

(5) If r=1 correlation is perfectly positive.

(6) If r=–1 correlation is perfectly negative.

(7) The closer the value of r is to 1 or -1, the higher the relationship between the variables.

(8) Correlation cofficient is the geometric mean between the two regression coefficients. *i.e.*,

$$r = \sqrt{b_{yx} \times b_{xy}}$$

Problem 1 : Find out coefficient of correlation between height of father and height of son from the following data and also calculate its Probable Error.

Height of father in cms. 165 166 167 167 168 169 170 172
Height of son in cms. 167 168 165 168 172 172 169
171

Solution :

x	dx=55	dx²	y	dy=13	dy²	dxdy
165	-5	25	167	-5	25	25
166	-4	16	168	-4	16	16
167	-3	09	165	-7	49	21
167	-3	09	168	-4	16	12
168	-2	04	172	0	0	0
169	-1	01	172	0	0	0
0	0	169	-3	9	0	0
172	2	04	171	–1	1	-2
N=8	Σdx=-16	Σdx^2=68	N=8	Σdy=–24	Σdy^2=116	$\Sigma dxdy$=72

Then by applying Karl Pearson's coefficient of correlation

$$r = \frac{n \times \Sigma dxdy - (\Sigma dx \times \Sigma dy)}{\sqrt{n \times \Sigma dx^2 - (\Sigma dx)^2} \times \sqrt{n \times \Sigma dy^2 - (\Sigma dy)^2}}$$

$$r = \frac{8 \times 72 - (-16 \times -24)}{\sqrt{8 \times 68 - (-16)^2} \times \sqrt{8 \times 116 - (-24)^2}}$$

$$r = \frac{576 - 384}{\sqrt{544 - 256} \times \sqrt{928 - 576}}$$

$$r = \frac{192}{\sqrt{288} \times \sqrt{352}}$$

Then by applying log-table

$$r = \text{Antilog of}\left[\log A - \frac{\log B + \log C}{2}\right]$$

$$r = \text{Antilog of}\left[\log 1920 - \frac{\log 288 + \log 352}{2}\right]$$

$$r = \text{Antilog of}\left[2.2833 - \frac{2.4594 + 12.5465}{2}\right]$$

$$r = \text{Antilog of}\left[2.2833 - \frac{5.0059}{2}\right]$$

$$r = \text{Antilog of}\left[2.2833 - 2.50295\right]$$

$$r = \text{Antilog of}\left[\bar{1}.78035\right]$$

$$r = 0.6030$$

Calculate of P.E.

$$\text{P.E.} = 0.6745 \times \frac{1 - r^2}{\sqrt{n}}$$

$$\text{P.E.} = 0.6745 \times \frac{1 - (0.6030)^2}{\sqrt{8}}$$

$$\text{P.E.} = 0.6745 \times \frac{1 - 0.363609}{2.8284}$$

$$\text{P.E.} = 0.6745 \times \frac{0.636391}{2.8284}$$

$$\text{P.E.} = 0.6745 \times 0.22500$$

$$\text{P.E.} = 0.0.15176$$

Problem 2 : Calculate coefficient of correlation and its Probable Error.

X = 20	24	28	32	36
Y = 14	18	22	26	30

Solution :

x	dx=28	dx^2	y	dy=26	dy^2	dxdy
20	-8	64	14	-12	144	96
24	-4	16	18	-8	64	32
28	-0	0	22	-4	16	0
32	4	16	26	0	0	0
36	8	64	30	4	16	32
N=5	Σdx=-0	Σdx^2=160	N=5	Σdy=-20	Σdy^2=240	$\Sigma dxdy$=160

Then by applying Karl Pearson's coefficient of correlation

$$r = \frac{n \times \Sigma dxdy - (\Sigma dx \times \Sigma dy)}{\sqrt{n \times \Sigma dx^2 - (\Sigma dx)^2} \times \sqrt{n \times \Sigma dy^2 - (\Sigma dy)^2}}$$

$$r = \frac{5 \times 160 - (-0 \times -20)}{\sqrt{5 \times 160 - (-0)^2} \times \sqrt{5 \times 240 - (-20)^2}}$$

$$r = \frac{800 - 0}{\sqrt{800 - 0} \times \sqrt{1200 - 400}}$$

$$r = \frac{800 - 0}{\sqrt{800} \times \sqrt{800}}$$

Then by applyi ng log-table

$$r = \text{Antilog of}\left[\log A - \frac{\log B + \log C}{2}\right]$$

$$r = \text{Antilog of}\left[\log 8000 - \frac{\log 800 + \log 800}{2}\right]$$

$$r = \text{Antilog of}\left[2.9031 - \frac{2.9031 + 2.9031}{2}\right]$$

$$r = \text{Antilog of}\left[2.9031 - \frac{5.8062}{2}\right]$$

$$r = \text{Antilog of}[2.9031 - 2.9031]$$

$$r = \text{Antilog of}[0.000]$$

$$r = 1.00$$

Regression

"Regression" means returing or stepping back to the average value. With the help of values of one variable (independent) we can establish most likely values of other variable (dependent). On the basis of two available correlated variables, we can forecast the future data or events or values.

In statistics, the term **"Regression"** means simply the "Average Relationship" We can predict or estimate the values of dependent variable from the given related values of independent variable with the help of a Regression Technique. The Measure of Regression studies the nature of correlationship to estimate the most probable values. It establishes a functional relationship between the 'Independent' and 'Dependent' variables.

The "Regression" succeeds the "Correlation". Once the correlationship between the two variables is established, the regression analysis proceeds with the estimation of probable values.

Nowdays, a modern statistician prefers to use the term "Regression" in the sense of "Estimation" which is an important statistical tool in economics and business. Estimation or prediction of economic activities is very essential planning. Estimating the relationship among the economic variables constitutes the essence of modern business management. That is why the term 'estimating line' is used instead of 'regression line' by the modern statisticians. Today, the term 'Regression' is used in a much broader sense to imply 'functional relationships.' It means the estimation or prediction of the unknown value of one variable from the known value of the other variable. The closer the relationship between the two variables, the greater the confidence may be placed in the estimates.

Regression Equation and Regression Coefficient

Algebric expressions of the regression lines are called the regression equation and also known as estimating equations. Because there are two regression lines, there are two regression equation. They are :

(i) Regression equation of y on x which determines the variations in the values of y for given changes in x.

(ii) Regression equation of x on y which determines the variations in the values of x for given chages in y.

Regression Equation of y on x

The regression equation of y on x is expressed as-

$$yc = a+bx \qquad \text{--------------(1)}$$

where, y_c = Value of y compoted from the relationship for a given x.

'a' and 'b' are constant and also known as the parameters of the line.

The parameter's determines the distances of line directly above of below the origin, while parameter 'b' determine the slope of the line i.e. the change y per unit change in x.

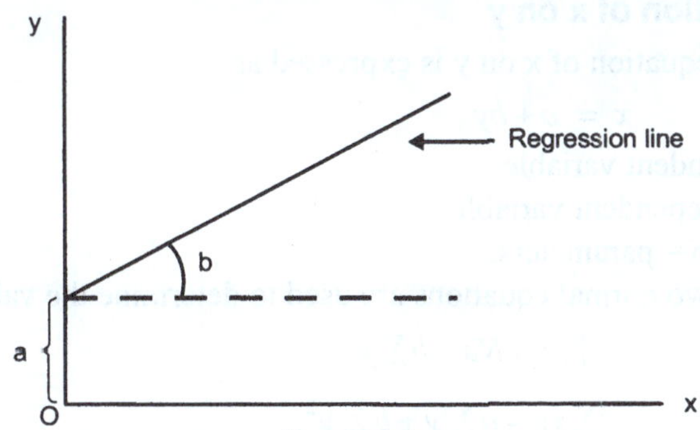

No correlation

The following two equations are solved simultaneously to find to the values of a and b.

$$\Sigma y = Na + b\Sigma x \quad ...(1)$$

$$\Sigma xy = a\Sigma x + b\Sigma x^2 \quad ...(2)$$

where = N = No. of paris of observations

These two equations usually called Normal Equation

From (1) Na = $Na = \sum y - b\sum x$

$$a = \frac{\sum y}{N} - \frac{b\sum x}{N} \quad a = (y - b\overline{x}) \quad ...(i)$$

From (2) $\sum xy = (y - b\overline{x})\sum x \ + b\sum x^2$

or $\sum xy = \overline{y}\sum x - b\overline{x} \ \sum x + b\sum x^2$

or $\sum xy = b\left[\sum x^2 - \overline{x}\sum x\right] + y\sum x$

or $b\left[\sum x^2 - \overline{x}\sum x\right] = \sum xy - \overline{y}\sum x$

$$b = \frac{\sum xy - \frac{\sum y}{N}\sum x}{\sum x^2 - \frac{\sum x}{N}\sum x} = \frac{N\sum xy - \sum x.\sum y}{N\sum x^2 - (\sum x)^2}$$

$$\therefore\ byx = \frac{N.\sum xy = (\sum x).(\sum y)}{N.\sum x^2 \ - \ (\sum x)^2} \quad ...(ii)$$

Where 'b' is called regression coefficient of y on x and is denoted by 'byx'

Now putting the value of 'a' and 'b' in equation (i). We get the required regression equation of y on x.

Thus $y = (\overline{y} - b\overline{x}) + bx$

$$(y - \overline{y}) = byx \ (x \ - \overline{x}) \quad ...(2)$$

Where $byx = \dfrac{N.\sum xy - (\sum x)(\sum y)}{N.\sum x^2 - (\sum x)^2}$

Regression Equation of x on y

The regression equation of x on y is expressed as

$$x = a + by \quad ...(1)$$

Where, x= dependent variable

y= independent variable

a and b = parameters.

The following two normal equations are used to determine the values of a and b.

$$\sum x = Na + b\sum y \quad ...(1)$$

$$\sum xy = a\sum y + b\sum y^2 \quad ...(2)$$

After solving equation (1) and (2) we get

$$a = (\bar{x} - b\bar{y}) \quad ...(i)$$

$$bxy = \frac{N\sum xy - (\sum x)(\sum y)}{N\sum y^2 - (\sum y)^2} \quad ...(ii)$$

Where 'b' is called regression coefficient of x on y and is written as bxy. Now, substituting the value of 'a' and 'b' in equation (1) we get the required regression lines of x on y.

Thus $x = (\bar{x} - b\bar{y}) + by$

or $(x - \bar{x}) = bxy(y - \bar{y})$ ...(2)

Method of Least Squares

The method of least suqares is a mathematical debvice, it is used for obtaining the equation of a curve which fits best a given set of observations. It is based on the assumptions that the sum of the square of differences between the estimated values and the actual observed values of the observation is minimise. The least square method is used to obtain the best fiting straight line to the given data and consists in selecting the best fitting straight line to the values of independent variable for dependents variable.

The method of least squares is used in obtaining the equations for fitting regression line.

Equation of a straight line is y = a + bx

Normal equation for obtaining the values of constant 'a' and 'b' are

(i) $\sum y = Na + b\sum x$ (ii) $\sum xy = a\sum x + b\sum x^2$

Where, x = Independent variable a= Constant

y = Dependent variable b=Constant

$\sum$ = Sign of summation N = No. of paris of observations.

The equation of the fitted straight line is y = a + bx

Where the values of 'a' and 'b' are obtained by solving the normal equation.

Problem 1 : Fit a straight line of y on x.

X =0	1	2	3	4	5	6
Y =2	1	3	2	4	3	5

Solution :

x	y	x^2	xy
0	2	0	0
1	1	1	1
2	3	4	6
3	2	9	6
4	4	16	16
5	3	25	15
6	5	36	30
$\sum x = 21$	$\sum y = 20$	$\sum x^2 = 91$	$\sum xy = 74$

The regression line of y on x is -

y = a + bx

The two normal equation which evaluate 'a' and 'b' are

$$\Sigma y = aN + b\Sigma x$$

$$\Sigma xy = a\Sigma x + b\Sigma x^2$$

Substituting the values of $\Sigma x, \Sigma x^2, \Sigma xy$ and N = 7 we get

20 = 7a + 21b ... (1)

74 = 21a + 91b ... (2)

Multiply equation (1) by 3 we get

60 = 21a + 91b ... (3)

Substracting (3) from (2) we have

14 = 28b ⇒ b=0.5

Putting b=0.5 in (1) we get

20 = 7a + 21b

20 = 7a + 21 (0.5)

20 = 7a + 10.5

20 = 10.5 = 7a

9.5 = 7a

$$\frac{9.5}{7} = a$$

1.35 = a

Substituting the value of a = 1.357 and b=0.5 in equation (A) we get

y = a + bx

y = 1.35 + 0.5x

Which is the line of best fit by least square method.

Problem 2 : Fit a straight line by least square method

X	=	0	1	2	3	4
Y	=	1.	1.8	3.3	4.5	6.3

x	y	x^2	xy
0	1.0	0	0
1	1.8	1	1.8
2	3.3	4	6.6
3	4.5	9	13.5
4	6.3	16	25.2
$\Sigma x = 10$	$\Sigma y = 16.9$	$\Sigma x^2 = 30$	$\Sigma xy = 47.1$

The equation of the line of fit is

y = a + bx ... (A)

Where a and b are constant

$$\sum y = Na + b\sum x$$

$$\sum xy = a\sum x + b\sum x^2$$

16.9 = 5a + 10b ... (1)

47.1 = 10a + 16.9b ... (2)

Multiply (1) by (2) we get

33.8 = 10a + 20b ... (3)

Substracting (2) from (3) we get

-13.3 = 3.1b ⇒ b = –4.29

Substituting b = -4.29 in (1) we get

16.9 = 5a – 42.9

or 5a = 59.8 ⇒ a=11.96

Substituting the values of | a = 11.96, b = –4.29 | in (A) we get the Eqn. of line as

y = 11.96 - 4.29x

Hence, the line of best fit is

y = -4.29x + 11.96

In Regression Analysis there are four important methods or types :

(1) When Two Series will be given

(2) When Arithmetic Mean and Standard-Deviation will be given

(3) When Equation will be given

(4) When Bivariate Frequency Distribution or Data will be given in the table

When two series will be given Regression Coefficient of -

x on y

$$Bxy = \frac{n.\Sigma dxdy - (\Sigma dx.\Sigma dy)}{n.\Sigma dy^2 - (\Sigma y)^2}$$

y on x

$$Byx = \frac{n.\Sigma dxdy - (\Sigma dx.\Sigma dy)}{n.\Sigma dx^2 - (\Sigma x)^2}$$

Ax and Ay

Ax = Mean of x series

$$Ax = x + \frac{\sum dx}{n}$$

Ay = Mean of y series

$$Ay = y + \frac{\sum dy}{n}$$

Regression Equation of

x on y

$$x - Ax = Bxy(y - Ay)$$

y on x

$$y - Ay = Byx(x - Ax)$$

Coefficient of correlation (r) = $\sqrt{Bxy.Byx}$

Note : If regression coefficient both are in mines (-) or any of them is in (-) then answer of coefficient of correlation is always (-)

Problem 3 : Following data given the various valus of two variables

x = 42 44 58 55 89 98 66

y = 50 49 53 58 64 76 58

Find out (i) Coefficicent of correlation

(ii) Two regression equation/limes of regression

(iii) Most likely values of x when y = 34

(iv) Most likely values of y when x = 47

Solution :

x	dx=58	dx^2	y	dy=26	dy^2	dxdy
42	–16	256	50	–8	64	128
44	–14	196	49	–9	81	126
58	–0	0	53	–5	25	0
55	–3	9	58	0	0	0
89	31	961	64	6	36	186
98	40	1600	76	18	324	720
66	8	64	58	0	0	0
N=7	$\sum dx$=-46	$\sum dx^2$=3086	N=7	$\sum dy$=-2	$\sum dy^2$=530	$\sum dxdy$=1160

(1) Regression Coefficient of x on y -

$$Bxy = \frac{n.\sum dxdy \;-\; (\sum dx.\sum dy)}{n.\sum dy^2 (\sum y)^2}$$

$$Bxy = \frac{7\times1160-(46\times2)}{7\times530-(2)^2}$$

$$Bxy = \frac{8120-92}{3710-4}$$

$$Bxy = \frac{8028}{3706}$$

$$Bxy = 2.166$$

(2) Regression Coefficient of y on x -

$$Byx = \frac{n.\sum dxdy-(\sum dx.\sum dy)}{n.\sum dx^2 (\sum x)^2}$$

$$Bxy = \frac{7\times1160-(46\times2)}{7\times3086-(46)^2}$$

$$Bxy = \frac{8120 - 92}{21602 - 2116}$$

$$Bxy = \frac{8028}{3706}$$

$$Bxy = 0.412$$

(2) Ax and Ay

X on y

$$Ax = x + \frac{\sum dx}{n}$$

$$Ax = 58 + \frac{46}{7}$$

$$Ax = 58 + 6.571$$

$$Ax = 64.571$$

y on x

$$Ay = y + \frac{\sum dy}{n}$$

$$Ay = 58 + \frac{2}{7}$$

$$Ay = 58 + 0.286$$

$$Ay = 58.286$$

(3) Regression Equation of

X on y

$x–Ax=Bxy(y–Ay)$

$x–64.571=2.166y(y–58.286)$

$x–64.571=2.166y–126.247$

$x=2.166y–126.247+64.571$

$x=2.166y–61.6676$

y on x

$y–Ay=Byx\ (x–Ax)$

$y–58.286=0.412(x–64–571)$

$y–58.286=0.412x–26.603$

$y=0.412x–26.603+58.286$

$y=0.412x+31.683$

(4) Calculate of x when y = 34

$x=2.166y-61.676$

$x=2.166x34–61.676$

$x=73.644–61.676$

$x=11.968$

(5) Calculate of x when y = 47

$y=0.412x+31.683$

$y=0.412x47+31.683$

$y=19.364+31.683$

$y=51.047$

(6) Coefficient of correlation $r = \sqrt{Bxy.Byx}$

$$r = \sqrt{2.166 \times 0.412}$$

$$r = \sqrt{0.892392}$$

$$r = 0.94$$

Problem 4 : Fit the regression equation of x on y and on x from the following data.

x=	10	20	30	40	50	60
y=	15	5	10	25	30	40

obtain the estimate of y when x = 22

Solution :

x	dx=30	dx^2	y	dy=25	dy^2	dxdy
10	–20	400	15	–10	100	200
20	–10	100	5	–20	400	200
30	0	0	10	–5	225	0
40	10	100	25	0	0	0
50	20	400	30	5	25	100
60	30	900	40	15	225	450
N=6	Σdx =–30	Σdx^2 =1900	N=6	$\Sigma dy = -25$	Σdy^2 =975	$\Sigma dxdy$ =950

(1) Regression Coefficient of x on y -

$$Bxy = \frac{n.\Sigma dxdy - (\Sigma dx.\Sigma dy)}{n.\Sigma dy^2 (\Sigma y)^2}$$

$$Bxy = \frac{6 \times 950 - (30 \times \ -25)}{6 \times 975 - (-25)^2}$$

$$Bxy = \frac{5700 \ - \ (-750)}{5850 \ - \ 625}$$

$$Bxy = \frac{5700 + 750}{5225}$$

$$Bxy = \frac{6450}{5225}$$

$$Bxy = 1.234$$

(2) Regression Coefficient of y on x

$$Byx = \frac{n.\Sigma dxdy \ - \ (\Sigma dx.\Sigma dy)}{n.\Sigma dx^2 (\Sigma x)^2}$$

$$Byx = \frac{6 \times 950 - (30 \times -25)}{6 \times 1900 - (30)^2}$$

$$Bxy = \frac{5700 - (-750)}{11400 - 900}$$

$$Byx = \frac{5700 + 750}{10500}$$

$$Byx = \frac{6450}{10500}$$

$$Byx = 0.614$$

(2) Ax and Ay

X on y

$$Ax = x + \frac{\sum dx}{n}$$

$$Ax = 30 + \frac{30}{6}$$

$$Ax = 30 + 5$$

$$Ax = 35$$

y on x

$$Ay = y + \frac{\sum dy}{n}$$

$$Ay = 25 + \frac{25}{6}$$

$$Ay = 25 - 4.167$$

$$Ay = 20.833$$

(3) Regression Equation of

X on y

x-Ax=Bxy(y-Ay)

x-35=1.234(y-20.833)

x-35=1.234y-25.708

x=1.234y-25.708

x=1.234y-25.708+35

x=1.234y+9.292

y on x

y-Ay=Byx (x-Ax)

y-20.833=0.614 (x-35)

y-20.833-0.614x-21.49

y=0.614x-21.49+20.833

y=0.614x-0657

(4) Calculate of x when y = 22

x=0.614x-0.657

x=0.614(22)-0.657

x=13.508-657

x=12.851

(5) Coefficient of correlation $r = \sqrt{Bxy.Byx}$

$$r = \sqrt{1.234 \times 0.614}$$

$$r = \sqrt{0.557676}$$

$$r = 0.87$$

Problem 5 : Formulate the appropriate regression equation from the following data :

X (Independent) :	2	4	5	6	8	11
Y (Dependent) :	18	12	10	8	7	5

Solution :

Since, Y is dependent variable, the approximate regression equation will be of Y on X which is computed as follows :

Computation of the Regression Equation of Y on X
(by the method of deviation from Mean)

X	Y	x (X–6)	y (Y–10)	xy	x^2	
2	18	–4	8	–32	16	
4	12	–2	2	–4	4	
5	10	–1	0	0	1	
6	8	0	–2	0	0	
8	7	2	–3	–6	4	
11	5	5	–5	–25	25	
$\Sigma X = 36$	$\Sigma X = 60$	-	-	$\Sigma xy = -67$	$\Sigma x^2 = 50$	N = 6

$$\bar{X} = \frac{\Sigma X}{N} = \frac{36}{6} = 6 \text{ and } \bar{Y} = \frac{\Sigma Y}{N} = \frac{60}{6} = 10$$

By the formula of regression equation of Y on X we have,

$$Y = \bar{Y} + \frac{\Sigma xy}{\Sigma x^2}(X - \bar{X})$$

$$= 10 + \frac{-67}{50}(X - 6)$$

$$= 10 + \frac{402}{50} - 1.34X = 10 + 8.04 - 1.34X$$

$$Y = 18.04 - 1.34X$$

QUESTIONS

1. Processing of data implies editing, coding, classfication and tabulation. Describe in brief these operations in the context of research study.
2. Processing of data implies cditing, coding, classification and tabulation." Describe in brief these four operations pointing out the significance of each in context of research study.
3. Why tabulation is considered essential in a research study? Narrate the characteristics of a good table.
4. Write a brief note on different types of analysis of data pointing out the significance of each.

■■■

CHAPTER

7 HYPOTHESES

After going through this chapter, you will be able to know :

- Types of Hypotheses
- Sources of Hypotheses
- Characteristics of Hypotheses
- Contents of Good Hypotheses
- Hypothesis Testing
- Advantages of Hypotheses
- Process of Hypotheses Testing
- Procedure for Testing of Hypotheses
- Steps in Hypotheses Testing
- Hypothesis Testing for the equality of two variances

Hypothesis is the proposed assumption, explaination, supposition or solutions to be proved or disproved.

It is considered as main instrument in research. It stands for the midpoint in the research.

Hypothesis is productive statement capable of being tested by scientific methods that are related to independent variable to some dependent variable.

If hypothesis is not formulated researcher cannot progress effectively. The main task in research is to test its record with facts. If hypothesis is proved the solution can be formed and if it is not proved then alternative hypotheses needs to be formulated and tested.

So, with hypothesis formulated it will help us to decide the type of data required to be collected. To answer the research question and the method in which the data should be organized effectively can be decided by proper hypothesis only.

The researcher makes queries or assumptions to solve the problem or guide further research, which is called hypotheses Hypothesis is provisional formulation of research problem. Based on these assumptions researcher anticipates certain logical consequences to be observed in research objectives.

The important function in research is formulation of hypothesis. The entire research activity is directed towards making of hypothesis. Research can begin with well formulated hypothesis or it may be the end product in research work.

Hypothesis gives us guidelines for an investigation to be carried out on the basis of previous available information. In absence of this, researcher will collect unrequired data and may eliminate required one. Thus hypothesis is an assumption which can be put to test to decide its validity.

Types of Hypotheses

The classifications of hypothesis so made on the basis of the level of abstraction are three level of abstraction for classification of hypothesis from lowest level to the highest level.

(a) At the lowest level of abstraction are the hypothesis which are at the state of existence of certain empirical uniformities. This is also called as common sense proposition without any justification. For eg: In rural India the child marriage is very common and they get married at the age group of 10-15 yrs.

(b) At relatively higher level of abstraction are the hypotheses which are concerned with complex ideal type.

(c) Last class of the hypothesis is the highest level of abstraction. It is concerned with the relation amongst the analytical variables.

Though various types of abstraction are explained here but none of these is described as superior to other one. Each type of hypothesis has its own importance.

Sources of Hypotheses

There are various sources of hypothesis based on which hypothesis can be developed. Some of the main sources are explained here:

1. Life history of a person decides his perception and this factor will direct researcher to form certain hypothesis. Individual intention is a source of hypothesis.
2. Similarities are often considered as source of valuable hypothesis. These similarities suffer with certain limitations, but they provide insights for formulation of hypothesis.
3. Hypothesis also arises form the findings of others studies. These studies which are conducted earlier provide the base for the new studies to be conducted.
4. The theoretical concept with logical deduction can be a source of hypothesis. The researcher may identify the variables which influence the status of concept. Hypothesis may be derived as deduction from the theoretical system.
5. The value orientation of the culture may provide the base for the hypothesis. Even personal happiness which is also a matter of study can be a source of hypothesis.

As discussed above the source of hypothesis is a combination of imagination and past experience of researcher.

Characteristics of Hypotheses

Hypothesis must satisfy the following criteria.

1. Hypothesis should be stated in such a manner by which inferences can be drawn and it can be tested by observation in the field.
2. Hypothesis should be clear, otherwise reliable inferences cannot bee drawn.
3. The hypothesis must be specific. The generality should be avoided and narrower hypothesis are more testable.
4. The hypothesis should be related to theory or theoretical orientation.
5. Hypothesis should be related to available techniques.
6. Hypothesis should be rational so that it can state the relationship between variables.
7. Hypothesis could explain the facts. It must explain what it claims and should have empirical references.
8. Hypothesis should be such, which can be tested within reasonable time.

Contents of Hypothesis

Research problem ultimately reduces to the question, which one of the alternatives is the effective one. The efficient among the alternative is always subject to condition. The statement of these acceptance of conditions are the hypothesis.

It is necessary for researcher to have resource survey which will bring alternative means, applied to research problem. Researcher must able to decide which alternative is most efficient in relation to criteria. The alternative hypotheses which researcher wants to formulate are statement of conditions for each alternative means have to be efficient.

Hypothesis Testing

In Business research and social science research, different approaches are used to study variety of issues. This type of research may be formal or informal, all research begins with generalized idea in form of hypotheses.

A research question is usually there. In the beginning research effort are made for area of study or it may take form of question about relationship between two or more variable. For example - do good working conditions improve employee productivity or another question might be now working conditions influence the employees' work.

Hypothesis differ from research question, it is specific and makes prediction about relationship between two or more variables. The difference between research question and hypotheses is that hypotheses predictics an outcome. For example- hypotheses may state positive relationship between working conditions and productivity of employee.

Advantages of hypotheses

1. It decides direction to research effort.
2. It make researcher to make clear the object of the research.
3. It also decide which variable will be considered for study which will not be considered for study.
4. It makes researcher to specify operational definition of the variables of interest.

The skill of researcher is an important factor on which usefulness of hypotheses is based. Hypotheses being basis of research study it should be developed with thoughtful consideration. A good hypotheses should have following contents.

Contents of Hypotheses

1. It should have logical consistency.
2. It must provide good rational for stepping away from main stream.
3. It must be able to test. Hypotheses is meaningful only when we design to conduct research.
4. It should be stated in simple and clear term to avoid confusion.

Process of Hypotheses Testing

Hypotheses testing is a systematic method. It is used to evaluate the data collected. This serve as aid in the process of decision making. The testing of hypotheses conducted through several steps which are given below.

1. State the hypotheses of interest.
2. Determine the appropriate test statistic.
3. Specify the level of statistical significance

4. Determine the decision rule for rejecting or not rejecting null hypotheses.
5. Collect the data and perform the needed calculations.
6. Decide to reject or not to reject the null hypotheses.

In order to provide more details on the above steps in the process of hypotheses testing each of test will be explained here with suitable example to make steps easy to understand.

1. Stating the hypotheses

In statistical analysis of any research study it includes at least two hypotheses, one is null hypotheses and another one is alternative hypotheses.

The hypotheses being tested is referred as the null hypotheses. and it is designated as H_0. It is also referred as hypotheses of difference. It should include a statement which has to be proved wrong.

The alternative hypotheses presents the alternative to null hypotheses. It includes the statement of inequality. The null hypotheses and alternative hypotheses are -complimentary.

The null hypotheses is the statement that is believed to be correct through analysis, which is based on this null hypotheses. For example- the null hypotheses might state the average age for entering management institute is 20 years. So average age for institute entry = 20 years.

If the data collected and analysed indicate the average entry age is greater than or less than 20 years the null hypotheses is rejected.

In this case alternative hypotheses could be stated in following ways.

a. The average age and entry to institute is not 20 years (=20)

b. the average age of entry to institute is less than 20 years (<2D)

c the average age of entry to institute is greater than 20 years (> 20)

The choice of which alternative hypotheses is to he used is decided as per the objective of the research study. The preceding b and c examples of above alternative hypotheses involve the use of a one tailed statistical test become as the direction of greater than or less than or vice versa.

Both (the alternative and the null hypotheses) must be decided before data collection. Both can be formulated to decide an expected conclusions of the study.

It is alternative hypotheses in which desired conclusions can be stated., provided null hypotheses include statement of equality.

For example an investigation is conducted to explore the effects of amount of study on marks in examination.

The investigator believes that study for longer (6) hours perform better than short (3) hours. In this case hypotheses might be–

The average examination marks of the students who study for 6 hours are (a) equal to the (=) average marks by the students who study for 3 hours. (b) less than the (<).

As a result of the statistical analysis the null hypotheses can be rejected or not rejected. It means null hypotheses can not be accepted.

If the null hypotheses is rejected, alternative hypotheses can be accepted however,

If the null hypotheses is not rejected than we can not conclude null hypotheses is true.

The rational believed this is that

a. evidence that support the hypotheses is not conclusive.

b. evidence that negates the hypotheses is damages hypotheses.

The analysis of study time and examination results of marks give an example. If the result of the study indicate that examination result (marks) of the students.

a. Who study 6 hours are better than the students who study for 3 hours. Therefore null hypothesis can be rejected because in this case null is not true.
b. Who study for 6 hours are not better than the student who study for 3 *hours* than null hypotheses can not he rejected

It can not he concluded that the null hypotheses is accepted because the results are only one set of marks comparisons.

Just because null hypotheses his true in one situation does not mean it is always true.

If the researcher sets up two hypotheses, there must be at least one point of agreement among them and one point of variance or disagreement. These alternative hypotheses my be represented symbolically as under :

H1 - MN1
H2 - MN2
H3 - MN3
H4 - MN4

The alternative hypotheses should encompass all possible outcomes of research, that is, they should be exhaustive with respect to the points of disagreement which will be tested. Secondly, of course, the hypotheses should be mutually exclusive. Failing these two requirements, research will not indicate which one course of action or solution should be selected from among the entire range of possibilities represented by the alternative hypotheses.

A very effective way of assuring ourselves that the hypotheses are mutually exclusive and jointly exhaustive of the universe of possibilities is to use the logical technique known as the "Boolean expansion". Suppose we have one common point of agreement (M) among the alternative hypotheses and three points of disagreement (e.g., N, O and P), then the alternative hypotheses according to the requirements of exhaustiveness and mutual exclusion could be presented as under. The common point of agreement among these hypotheses may be the examination score under specific conditions. Thus, M = Examination score. The points of difference may be N = more than x; N' = less than x; similarly O = more than y, O' = less than y and P = more than z, P' = less than z.

(Read H4 as Examination score is more than x and y but less than z).

H1 - MNOP	H5 - MN'O'P
H2 - MN'OP	H6 - MN'OP'
H3 - MNO'P	H7 - MNO'P'
H4 - MNOP'	H8 - MN'O'P'

In general, if there are n points of disagreement, there will be 2 n (2 x 2 x 2 x... n times) alternative hypotheses in an exclusive classification. Only one of them can be true and must be true.

In a research involving more than two hypotheses, it is advisable to formulate the points of disagreement symbolically in a manner indicated to facilitate the construction of hypotheses. Intuition is often not a satisfactory guide.

It was suggested earlier that ideally there should be one hypothesis for each alternative course of action. Such a problem is one which involves estimation, e.g., estimating an optimum number of workers for a production unit - 100, 250, 300 etc. The selection of the most efficient course of action depends on an estimate of the value of a critical variable (i.e., the exact number of workders). In such cases, it is not economical to formulate explicitly each alternative course of action and to associate a hypothesis with each. We can use only a short-hand formulation. The alternate hypothesis can simply be stated as: "K workers are needed" and the problem of research is to estimate this K. Now, since the estimate of the value of any variable is subject to error, it is advisable to express the estimate as a range of values rather than a single value, e.g., 300 + 50 workders (250 to 350) are needed.

We may now introduce the readers to the concept of null hypothesis. The alternative hypotheses and the null hypotheses together constitute the framework for the statistical testing of hypotheses. Fuller significance of null hypothesis can be grasped by the students at a later stage. Suffice it here to say that the null hypothesis in its simplest form asserts that there is no difference between two populations in respect of some property and that the difference found between the samples drawn from these populations is only accidental and unimportant. H.M. Garrett remarks: "the null hypothesis is akin to the legal principle that a man is innocent until he is proved guilty. It constitutes a challenge and the function of a (research) is to give facts a chance to refute this challenge."[14] A null hypothesis in its other form may assert that the results found in research do not differe significantly from the results to be expected on a probability basis or as stipulated in terms of a certain theory. An alternative hypothesis may state, for instance, that:

H1 : The females commit suicide often than the males.

The null hypothesis (Ho) may be stated as:

Ho : The females and males do not differ in respect of the rate of suicide.

2. Determining Appropriate test statistic

The appropriate test statistic, which is to be used in statistical hypotheses testing, is based

on various characteristics of sample population of interest including sample size and distribution.

The test statistic can assume many numerical values. As the value of test statistic has significant on decision one must use the appropriate statistic in order to obtain meaningful results.

The formula to be used while testing population means is :

Z = test statistic

X = mean of sample

μ = mean of population

σ = standard deviation

n = number of sample

3. The Significance Level

As already explain, null hypotheses can be rejected or fail to reject null hypotheses. A null hypotheses that is rejected may in really be true or false.

A null hypotheses that fails to be rejected may in reality be true or false. The outcome that a researcher desires is to reject a false null hypotheses or fail to reject a true null hypotheses. However there is always possibility of rejecting a true hypotheses or failing to reject false hypotheses.

Type I and Type II Error

Type I error is rejecting a null hypotheses that is true.

Type II error is failing to rejected a false null hypotheses

Table for type I and type II Error

	DECISION	
	ACCEPT H_0	REJECT H_0
H_0 (TRUE)	CORRECT DECISION	TYPE I ERROR
H_0 (FALSE)	TYPE II ERROR	CORRECT DECISION

The probability of committing a Type I error is termed as A and Type II error is termed as B.

As the value of A increases the probability of committing Type I error increases.

As the value of B increases the probability of committing Type iI error increases.

Researcher would like to decrease the probability of committing both types of errors. The reduction of A, result in the increase of B and vica-versa.

The best way to reduce the probability of decreasing both types of errors is to increase the size sample. The level of significance is probability of committing. Type I error. Now it is necessary that one must specify the level of significance before data is collected. It means probability of committing Type I error regarding rejecting a true null hypotheses.

There is opposite relationship, between avoiding of Type I error and selecting value of A in the mind of researcher. So if not making error is important than low probability of making error is sought. The more the desire not to reject a true null hypotheses lower the select value of A.

In theoretical principles the value of A is in between 0-I. However mostly value used in business

research are 05. 01 and 001 which respectively correspond to level of 95. 99 and 99.9 p.c. therefore there is possibility of Type I error is not being made. By selecting higher level of significance it will make strong statistical evidence to ever reject the null hypotheses.

4. Decision Rule

Before collection and analyses of data it is necessary to decide under which conditions the null hypotheses will be rejected or fail to be rejected.

The decision rule can be stated in terms of computed test statistic, or in probabilitistic terms. The same decision will he applicable any of the method so selected.

5. Data Collection and Calculation Performance

In research process at early stage method of data collection is decided. Once the research problem is decided than immediately decision in respect of type and source of data should be taken. It must clear the fact that, which type of data will be needed for the purpose of the study and now researcher has a plan to collect required data.

The decision will provide base for processing and analysing of data. It is advisable to make use of approved methods of research for collecting and analysing of data.

6. Decision on Null Hypotheses

The decision regarding null hypotheses in an important step in the process of the decision rule.

Under the said decision rule one has to reject or fail to reject the null hypotheses.

If null hypotheses is rejected than alternative hypotheses can be accepted. If one fails to reject null hypotheses one can only suggest that null hypotheses may be true.

7. Two failed and one failed tests

In the case of testing of hypotheses, above referred both terms are quite important and they must be clearly understood. A two failed test rejects the null hypotheses.

a. If sample mean is significantly
b. higher or lower than the
c. hypothesized value of mean of the population.
d. Such a test is appropriate, when the null hypotheses is some specified value and the
e. alternate hypotheses is a value not equal to the specified value and the alternative hypotheses is value not equal to the specified value of null hypotheses. There are certain situations when only.
f. one failed test is considered appropriate. One failed test is used when we want to test weather the population mean is either lower than or higher than some hypothesized value.

Procedure for Testing of Hypotheses

Testing of hypotheses means to decide the validity of the hypotheses on the basis of the data collected by researcher. In testing procedure we have to decide weather null hypotheses is accepted or not accepted.

This requirement is conducted through several steps between the course of two action i.e. rejection or acceptance of null hypotheses. The steps involved in the testing of hypotheses are given below :

1. Setting up of Hypotheses

This step consist of hypotheses setting. In this step formal statement in relation to hypotheses in made. In traditional practice instead of one, two hypotheses are set. In case it one hypotheses is rejected than other hypotheses is accepted. Hypotheses should be clearly stated in respect of the nature of the research problem. There hypotheses are

A. Null hypotheses and
B. Alternative hypotheses.

Acceptance or rejection of hypotheses is based on the sampling information. Any sample which we draw from the population will vary from it therefore it is necessary to judge whether there differences are statistically significant or insignificant.

The formulation of hypotheses is an important step which must be accomplished and necessary
care should be taken as per the requirements and object of the research problem under construction.

This should also specify the whether one failed or two failed test will be used.

2. Selecting Statistical Technique

In this stage we will make selection of statistical technique which are going to he used. There arc various statistical test which arc being used in testing of hypotheses. There tests are Z-test, T-test. F-test and X2.

It is the jobs of the researcher to make proper selection of the test.

A - Z - test is used when hypotheses is related to a large sample (30 or more)

b - 't' test is used when hypotheses is related to small sample (less than 30)

The selection of test will be dependent on various considerations like, variables involved, sample size, type of data and whether samples are related or independent.

3. Selecting level of Significance.

This stage consist of making selection of desired level of significance. The researcher should specify level of significance because testing of hypotheses is based on 'pre-determined level of significance. The rejection or retention of hypothesis by the researcher is also based on the significance level.

The level of significance is generally expressed in percentage form such as 5% or 1%, 5% level of significance is accepted by the researcher, it means he will be making wrong decision about 5% of time. In case if hypotheses is reject at this level of 5% he will be entering risk hypotheses rejection in 5 out of 100 occasions.

The following factors may affect the level of significance.

(i) the magnitude difference between sample means.

(ii) the size of sample

(iii) the validity of measurement within sample.

(iv) Whether the hypotheses is directional or non-directional.

The level of significance must be adequate in relation to the nature and object of the research study.

4. Determining Sampling distribution

The next step after deciding significance level in testing of hypotheses is to determine the appropriate sampling distribution. It is, normal distribution and 't'- distribution in which choice can be excised.

5. Selecting Sample and Value

In this step random sample is selected and appropriate value is computed., from the sample data relating to the test statistic by utilizing the relevant distribution.

6. Performance Computation

In this step calculation of performance is done. The calculation includes testing statistics and standard error.

A hypotheses is tested for following four possibilities ,that the hypotheses is -

a - true, but test leads to its rejection.

b - false, but test leads to its acceptance.

c - truc, but test leads to its acceptance.

d - false, but test leads to its rejection.

Out of the above four possibilities a and b lends to wrong decision.

In this case a lends to Type I error and, b lends to Type II error.

7. Statistical Decision

Thus is the last step in which we have to draw statistical decision involving the acceptance or rejection of hypotheses.

This will be dependent on whether the calculated value of the test falls in the region of acceptance or in the region of rejection at given significance level.

If hypotheses is tested at 5% level and observed set recorded the possibilities less than 5% level than we considered difference between hypothetical parameter and sample statistics is significant.

Steps in Hypothesis Testing

The following four steps are carried out in testing a hypothesis :

Step I : State the null hypothesis (H_0) and the alternative hypothesis (H_1).

Step II : Compute the value of the appropriate test statistic on which the decision to accept or reject H_0 depends.

Step III : Decide on the acceptance or rejection of H_0 on the basis of the value of the test statistic in Step II at a particular level of significance (α).

Step IV : Make appropriate conclusions based on Step III.

Different kinds of tests in this Chapter and in Chapters 4 and 5 have been carried out in these four steps.

Tests for Population Mean

Large Sample Test (n > 30)

The random variable χ is assumed to follow a normal distribution with mean μ and standard deviation σ, i.e. χ is N (μ, σ). The mean of a random sample of size n(> 30) is denoted by $\bar{\chi}$. It can be shown that $\bar{\chi}$ follows a normal distribution with mean μ and standard deviation $\sigma/\sqrt{n}$ .Therefore, $[(\bar{\chi} - \mu)\sqrt{n}]/\sigma$ is normally distributed with mean zero and standard deviation unity. If the null hypothesis $H_0 : (\mu = \mu_0)$ is true, then

$$Z = [(\bar{\chi} - \mu_0)\sqrt{n}]/\sigma \qquad (1)$$

is normally distributed with mean zero and standard deviation unity, i.e. z is N(0,1).

In order to test the null hypothesis $H_0 : (\mu = \mu_0)$ against an appropriate alternative hypothesis H_1 at 5% level of significance.

(i) accept $H_1 : (\mu < \mu_0)$ if the calculated value of z is less than (- 1.64)

(ii) accept $H_1 : (\mu > \mu_0)$ if the calculated value of z is greater than (+ 1.64)

(iii) accept $H_1 : (\mu \neq \mu_0)$ if the calculated value of z is less than (- 1.96) or greater than (+ 1.96)

For the alternative hypotheses $H_1 : (\mu < \mu_0)$, $H_1 : (\mu > \mu_0)$ and $H_1 : (\mu \neq \mu_0)$, the quantities (- 1.64), (+ 1.64) and ($\pm$ 1.96) respectively are called the 'critical values' of z at 5% level of significance. The corresponding critical values at 1% level of significance are (- 2.33), (+ 2.33) and ($\pm$ 2.58), respectively.

At 5% level of significance, the area under the normal distribution N(0, 1) to the left of (-1.64) for the alternative hypothesis $H_1 : (\mu < \mu_0)$ is called 'critical region'. Similarly, the critical region for the alternative hypothesis $H_1 : (\mu < \mu_0)$ is the area to the right of (+ 1.64). For the alternative hypothesis $H_1 : (\mu \neq \mu_0)$, the critical region is the sum of the areas to the left of (- 1.96) and right of (+ 1.96), each representing 2.5% of the area under the normal curve. The critical regions at 1% level of significance may similarly be defined. The null hypothesis H_0 is rejected whenever the test statistic z in (1) falls in the critical region, otherwise it is accepted. When z is exactly or approximately equal to the critical value, no decision should be made for the acceptance or rejection of H_0. Similarly, decision is suspended if z lies between its critical values at 5% and 1% levels of significance. In such cases, more data should be collected for calculating z afresh, so that it falls within or outside the critical region. Critical values of z at level of significance α for different types of alternative hypotheses are shown in figures

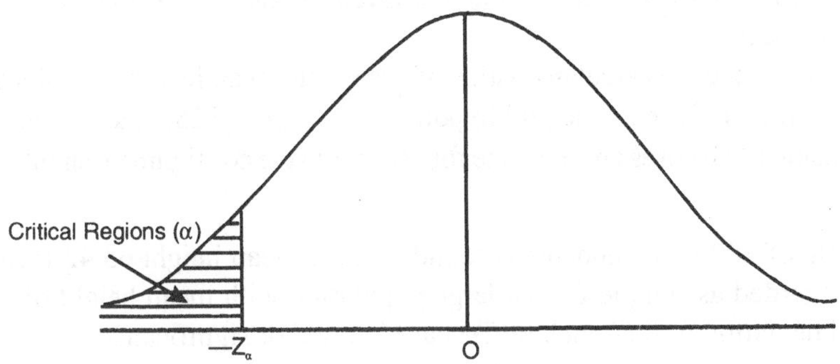

Critical Value of z at Level of Significance α for the Alternative Hypothesis $H_1 : (\mu < \mu_0)$

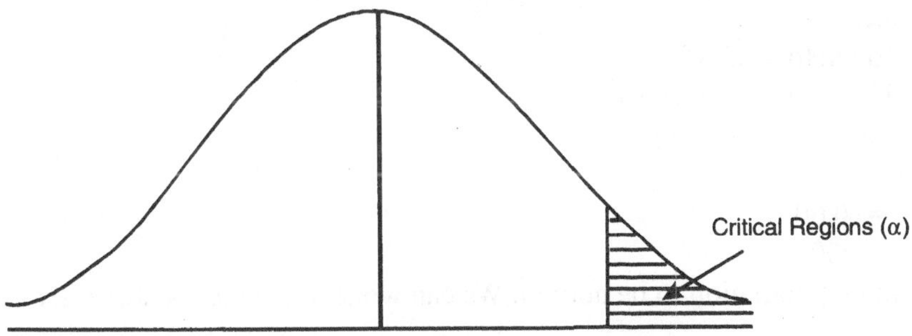

Critical Value of z at Level of Significance α for the Alternative Hypothesis $H_1 : (\mu > \mu_0)$

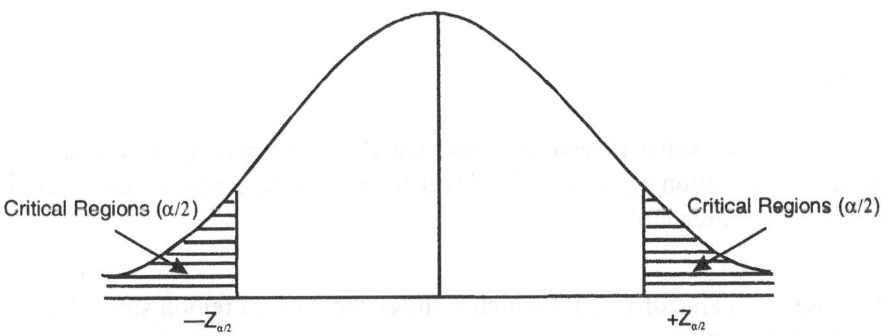

Critical Value of z at Level of Significance α for the Alternative Hypothesis $H_1 : (\mu \neq \mu_0)$

Problem 1

A particular drug is bought in bulk quantities by a hospital from a drug and pharmaceutical company. The mean dose is 125 cc. Excess doses are not harmful to the body. Insufficient doses, however, are not effective. Population standard deviation based on the analysis of a large mass of past data is known to be 3 cc. The mean of a random sample of 70 doses drawn from a large consignment supplied by the manufacturer is 124.5 cc. Should the consignment be accepted?

If μ denotes the mean dose, the null and alternative hypotheses are $H_0 : (\mu = 125)$ and $H_1 : (\mu < 125)$ respectively. The consignment is accepted if H_0 is accepted, otherwise the consignment is rejected. From (1),

$$Z = [(\bar{x} - \mu_0)\sqrt{n}]/\sigma = [(124.5 - 125.0)\sqrt{70})]/3$$

$$= (-0.5)(8.37)/3 = -1.395.$$

The critical values of z at 5% and 1% levels of significance are $z_{0.05} = -1.64$ and $z_{0.01} = -2.33$ respectively.

Decision : Since the calculated value of z is greater than its critical values at both 5% and 1% levels of significance, the null hypothesis $H_0 : (\mu = 125)$ is accepted.

Conclusion : The doses are adequate for use and so the consignment should be accepted.

Problem 2

A sample of 200 male students is found to have mean height 65.42 inches can it be reasonably regarded as sample from a large population with mean height of 65.36 inches and standard deviation of 0.60 inches" Test at 5% level of significance.

Solution

Taking null hypotheses that the mean heighs of the population is equal to 65.36 inches, we can while.

Ho : uHo = 65.36

H1 = uHL # 65.36

and in given formula as

x = 65.52

= 0.60

n = 200

Assuming population to be normal. We can work out the test statistic Z as under.

$$Z = \frac{X - Ho}{\sqrt{n}} = \frac{65.42 - 65.36}{0.60/\sqrt{200}}$$

$$= \frac{0.06}{0.042}$$

$$= 1.4$$

As H0\o is two sided in give question we shall be applying a two failed test for determining the rejection regions at 5% level to significance which comes to as under, using normal curve area table

R : L 3 I > 1.96

The observed value of Z is 1.4 which is the acceptance of region since R : IZ 1> 1.96 and thus HO is accepted.

We may conclude that the given sample (with mean height) 65.42) can be regarded to have been taken from a population with means highs 65.36 and standard deviation 0.60 at 5% level of significance.

Problem 3

AB Co focused on having stable workforce with little turnover. Co is in business from last 30 years and has more than 5000 employees.

The co has promoted the idea that its employees stay long time, and so it has included this information in co brochures The average turnover of our employee is 20 years.

Since co is not sure about this statement that it is true, a random survey 100 employees conducted and then average age turn to be 19 years with standard deviation of 27 years. *Can* ABCo continue to make its claim or does it need to make change?

1. State the hypotheses

   Ho = 20 years

   H I ≠ 20 years

2. Determine test statistic. Since we are testing population mean that is normally distributed the appropriate test statisticis.

3. Specify the significance level. Since the firm would like to keep its present message to new

   recruits it selects weak significant level (a = .05)

   Since this is a two failed test half of the alpha will be assigned to each tall of the distribution. In this situation the critical value of

   Z = + 1.96 and - 1.96

4. State the dicision rule. If the computed value of Z is greater than or equal to +1.96 or less than or equal to -1.96 the null hypotheses is rejected

5. Reject or fail to reject the null since 2.5 is greater than1.96 the null is rejected. The mean tenure is not 20 years. therefore AB Co needs to change its statement.

Problem 4

10 Industrial units of same size facing problem of labour turn- over. The mean distribution of annual turnover is 160 employees with standard deviation 37 employees.

Random sample of 5 industrial units taken which give mean of 150 employees turnover. Is the sample mean consistent with the population

Solution -

Considering null hypotheses of the population mean is 160 employee We can say

Ho = 160

HI ≠ 160

and other information is as under

x 150

37

n 5

N 10

Assuming normal populationtest statistic Z as under.

$$Z = \frac{\overline{X} - Ho}{\delta/\sqrt{n} \times \sqrt{(N-n)/(N-1)}}$$

$$Z = \frac{150-160}{\sqrt{37/\sqrt{5}} \times \sqrt{(10-5)/(10-1)}}$$

$$= \frac{-10}{(16.8)\ (.745)} = \frac{-10}{12.5} = -0.8$$

As Ho is two sided in the given question we shall apply a two tailed test for deciding the rejection region at 5% level of significance. Which comes to as under using normal curve area table.

R : I Z I > 1.96

The observed value of Z is - 0.8 which is in the acceptance region since R : IZ 1> 1.96 and thus Ho is accepted and we may conclude that the sample mean is consistent with population mean it means the population mean 160 is supported by sample results.

Hypothesis Testing for the equality of two variances

χ_1 and χ_2 are two normally distributed independent variables with the following population means and standard deviations :

μ_1 = Population mean of χ_1 (first population)

μ_2 = Population mean of χ_2 (second population)

σ_1^2 = Population variance of χ_1 (first population)

σ_2^2 = Population mean of χ_2 (second population)

Random samples of sizes n1 and n2 are drawn from the first and second populations respectively, the corresponding sample means being $\bar{\chi}_1$ and $\bar{\chi}_2$. Let χ_{1i} (i = 1, 2, ..., n_1) and $\bar{\chi}_{2j}$ (j = 1, 2, ..., n_2) denote the sample observations. The sample variances are,

$$S_1^2 = \Sigma\ (\chi_{1i} - \bar{\chi}_1)^2\ /\ (n_1\ -\ 1)$$

$$S_2^2 = \Sigma\ (\chi_{2i}\ - \bar{\chi}_2)^2 / (n_2 - 1)$$

The denominators are (n1 - 1) and (n2 - 1) in the expressions for S_1^2 and S_2^2 to make them unbiased estimators of σ_1^2 and σ_2^2 respectively. The concept of unbiasedness has been defined in Section 3.5.2.

The null hypotheis $H_0 : (\sigma_1^2 = \sigma_2^2)$ is required to be tested against one of the following three alternative hypotheses :

$H_1 : (\sigma_1^2 < \sigma_2^2)$ (one-sided 'less than type' or left-tailed alternative hypothesis)

$H_1 : (\sigma_1^2 > \sigma_2^2)$ (one-sided 'greater than type' or right-tailed alternative hypothesis)

$H_1 : (\sigma_1^2 \neq \sigma_2^2)$ (two-sided alternative hypothesis)

If the null hypothesis $H_0 : (\sigma_1^2 = \sigma_2^2)$ is true, then

$$F = F = S_1^2 / S_2^2$$

follows F distribution with degrees of freedom $(n_1 - 1), (n_2 - 1)$. F lies between 0 and , and its shape depends on the degrees of freedom. In the computation F, the larger of S_1^2 and S_2^2 should be in the numerator, so that F > 1. Critical values of F at 5% and 1% levels of significance are given in Appendices VI and VII respectively. A typical *F* distribution is shown in Figure

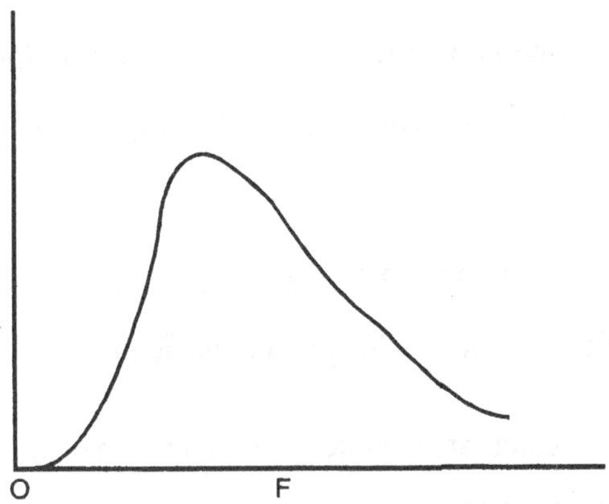

Typical F Distribution

In Figure $f(F)$ denotes the density function of F $(0 \leq F < \infty)$ and so $\int_0^{\alpha} f\ (F)\ dF = 1.$

Problem 5

The daily outputs of two similar machines in terms of number of units produced are as follows :

Daily Outputs of Two Similar Machines

Day	1	2	3	4	5	6	7	8	9	10	11	12
Machine I	20	16	26	27	23	22	18	24	25	19	-	-
Machine II	17	23	32	25	22	24	28	18	31	33	20	27

Is machine I more consistent than machine II in terms of daily output ?

Let x_{1i} (i = 1,2, ..., 10) x_{2j} (j = 1, 2, ..., 12) denote the sample observations from machines I and II respectively. If n_1 and n_2 are the sample sizes, $\bar{x}_1$ and $\bar{x}_2$ are the sample means and S_1^2 and S_2^2 are the sample variances of the daily outputs of machines I and II, respectively, then

$$\bar{x}_1 = \Sigma x_{1i} / n_1 = 220/10 = 22$$

$$\bar{x}_2 = \Sigma x_2 / n_2 = 300/12 = 25$$

$$S_1^2 = \Sigma (x_{1i} - \bar{x}_1)^2 / (n_1 - 1)$$
$$= [(20 - 22)^2 + (16 - 22)^2 + ... + (25 - 22)^2 + (19 - 22)^2] / 9$$
$$= 13.33$$

$$S_2^2 = \Sigma (x_{2i} - \bar{x}_2)^2 / (n_2 - 1)$$
$$= [(17 - 25)^2 + (23 - 25)^2 + ... + (20 - 25)^2 + (27 - 25)^2] / 11$$
$$= 28.55$$

If σ_1^2 and σ_2^2 are the population variances of daily outputs of machines I and II respectively, the null and alternative hypotheses are $H_0 : (\sigma_1^2 = \sigma_2^2)$ and $H_1 : (\sigma_1^2 < \sigma_2^2)$ or $H_1 : (\sigma_2^2 > \sigma_1^2)$. From (4)

$$F = S_2^2 / S_1^2 = 28.55 / 13.33 = 2.14 \text{ (Degrees of freedom = 11,9)}$$

Critical values of F at 5% and 1% levels of significance are $F_{0.05;\,11,9} = 3.105$ and $F_{0.01;\,11,9} = 5.185$ respectively.

Decision : Since the calculated value of F is less than its critical values, the null hypothesis $H_0 : (\sigma_1^2 = \sigma_2^2)$ is accepted.

Conclusion : There is no statistically significant difference in the consistency of the daily outputs of the two machines.

A company buys raw materials from two suppliers A and B. It is interested in ascertaining which of the two suppliers is more consistent with respect to a particular characteristic of raw material quality. A random sample of size 10 drawn from the lot of supplier A has a variance of 150 while that of size 15 drawn from the lot of supplier B has a variance of 120. Is it correct to conclude that supplier B is more consistent?

Let n_1 and n_2 denote the sizes of samples drawn from the lots of suppliers A and B, and S_1^2 and S_2^2 denote the corresponding sample variances. Here $n_1 = 10$, $n_2 = 15$, $S_1^2 = 150$ and $S_2^2 = 120$. If σ_1^2 and σ_2^2 denote the population variances, the null and alternative hypotheses are $H_0 : (\sigma_1^2 = \sigma_2^2)$ and $H_1 : (\sigma_1^2 < \sigma_2^2)$ respectively. From (4)

$$F = S_2^2 / S_1^2 = 150/120 = 1.25 \text{ (Degrees of freedom = 9,14)}$$

Critical values of F at 5% and 1% levelsof significance (degrees of freedom = 9, 14) are $F_{0.05;\,9,14} = 2.65$ and $F_{0.01;\,9.14} = 4.03$ respectively.

Decision : Since the calculated value of F is less than its critical values, the null hypothesis $H_0 : (\sigma_1^2 = \sigma_2^2)$ is accepted.

Conclusion : There is no statistically significant difference in the consistency of the qulity of raw materials purchased from suppl/iers A and B.

QUESTIONS

1. Explain the characteristics of hypothesis. Describe the procedure of hypothesis testing.
2. The procedure of testing hypothesis requires a researcher to adopt several steps. Describe in brief all such steps.
3. What do you mean by the power of a hypothesis test? How can it be messured? Describe ab illustrate by an example.
4. Define hypothesis. What are the types of hypothesis. Explain them with suitable examples.

■■■

CHAPTER

8 ANALYSIS OF VARIANCE

After going through this chapter, you will be able to know :

- One way annova
- Two way annova
- 'f' test
- One annova table
- Coding Method
- Short cut-Method
- Interpretation of data
- Necessary of Interpretation
- Techniques of Interpretation
- Preventive measures for Interpreation

A-nova'- Analysis of variance, it is a testing technique normally used in the social science research study.
where multiple - sample cases on involved.

As already explained earlier 'z' test and 't' test are used to judge the significance difference where two or samples are there.

Where more than two samples are required to he judged at the same time than ANOVA test can be used. By making use of this technique. We can work out the fact regarding sample drawn from the population having, same mean. This test is important in reference to situations where more than two variables are to compared.

AVOVA is a procedure for testing the difference in various data group for similarity. In this method variance is divided for the purpose of analysis. It is variance analyzing method where response in based on various factors referring to different source of variation.

In this context it is necessary to explain variance. It is important statistical measure and
is describe as the mean of squares of deviations
from the mean of given series of data. It is used measure of variation. It in known as standard deviation = Variance

Any number factors which influenced the dependent variable in the research study can he examined through Anova technique. Even beyond this examination it can also be conducted for differences amongst various categories within each factor.

If we take one factor and investigate the differences amongst' its various categories are called **one way anova**.

Where we investigate two or move factors at the same time it is known as **two way anova**.

Anova is to test for differences among means of population by examining amount of variation within each of the sample relates to the amount of variation between sample.

So two estimates of populations are made

1. Based on between sample variance.
2. Based on within sample variance.

These two estimates are compared with 'f' test.

$$\text{'F'} = \frac{1}{2}\ \frac{\text{between sample}}{\text{with in sample}}$$

This 'f' is to compared to 'f' limit for given degree of freedom. 'f' value equal or exceed to 'f' limit.

we may say that there are significant difference between sample means.

Technique of Anova related to

(i) One way Anova

(ii) Two way Anova

(1) One way Anova - In this technique only one factor is considered than observed, in this factor several possible sample can occur Now we determine if there are difference within, factor. The following steps are involved.

(a) Mean of each sample to be obtained

If $\bar{x}, \bar{x}_2, \bar{x}_3, \bar{x}_4, \bar{x}_5$

When 'k' samples are there calculate mean of sample mean as

$$\bar{x} = \frac{\bar{x}, \bar{x}_2, \bar{x}_3, \bar{x}_4, \bar{x}_5}{\text{No. of sample k}}$$

(b) Calculate deviations of sample mean from the mean of the sample mean Calculate the square of such deviation and multiply by the item number in sample and obtain total. SS - knows as sum of square of variance
= 1272 - 1200 = 72

(c) The result of step b divided by degree of freedom between sample or mean square (MS) between sample

$$\text{MS between} = \frac{\text{SS between}}{(k-1)}$$

(k-1) means df (degree of freedom) between samples.

(d) Obtain the deviations of values of sample items from respective means of the sample and calculate square of such deviation and total. This total in SS. (sum of square of variance within sample)

$$\text{SS with in} = \Sigma(xij - \bar{x}_1)^2 + \Sigma(xij - \bar{x}_2)$$

(e) divide the result of step 'd' by degree of freedom within sample and obtain mean square (MS) with in sample

$$\text{MS with in}\ \frac{\text{SS with in}}{(n-k)}$$

(n-k) is degree of freedom with sample n, is total number of items in all sample k is number of sample.

(f) Sum of square of deviations can also worked out by adding square of deviation when deviation for individual items in all sample are taken from mean of the sample . It can he explained as SS for total variance

$= \Sigma(xij - \bar{x})^2 \quad i = 1.2.3$

$j = 1.2.3$

This total should be result of b and c

SS for total variance — SS between + SS within
The degree of freedom for total variance — number of items in all sample - I (n-1)
The degree of freedom for between and → add up to degree total variance within

(n-1)=(k-1)+(n-k)

(g) Lastly 'f' ratio worked out as under

$$F\text{ ratio} = \frac{\text{MS between}}{\text{MS with in}}$$

Table for one way Anova.

In order to have easy understandability information in given in the formulas as above can also be presented in tabular form as under One way Anova Table.

Source Varation	Sum of source (S.S.)	Degree of freedom (df)	Mean square (MS) (SS/df) Varation to be used for f ratio	'F' Ratio
Between	$n_1(\bar{x}_1 - \bar{\bar{x}})^2 +$			
sample		(k-I)	$\frac{\text{ss between}}{(k-1)}$	$\frac{\text{MS between}}{\text{MS within}}$
Within Samples or	$\Sigma(xij - x_1)^2 +$			
Categories	$+\Sigma(xij - \bar{x}k)^2$ i =1.2.3.	(n- k)	$\frac{\text{SS with}}{(n-k)}$	
Total	$+\Sigma(xij - \bar{x}k)^2$	(n-1)		

I = 1, 2. J = 1, 2.

One-way anova short cut method.

As this method is very faverable to requirments therefore normally used in such cases when means of sample or mean and sample means are in to non-integer values.

The steps involved in the short cut method are given as below

(a) Work out $\Sigma\, xij$ by taking total value of individual items in all sample.

$\Sigma\, xij = i = 1.2.3.$

$j = 1.2.3.$ call as 'T'

(b) now work out correction factor

$$\text{Correction factor - } -\frac{(T)^2}{n}$$

(c) Calculate square of all items one by one - subtract correction from.
The total - result is sum of squares of total variance.

$$\Sigma x2ij - \frac{(T)^2}{n} i = 1.2.3.$$

Total SS j – 1.2.3.

(d) **obtain** square of each sample total (Tj)-' **devide** square value by number of item. **substract** correction factor from the from this total and now the result is sum of square of variance between sample.

$$\text{SS between} = \Sigma = \frac{(Tj)^2}{nj} - \frac{(T)^2}{nj} - \frac{(T)^2}{n} = j1.2.3.$$

j = different samples or categories

(e) The of square within sample can be found by subtrating the result of 'd' stop from result of 'c' step.

$$SS = \left[\Sigma x^2 ij - \frac{(T)2}{n}\right] - \left[\Sigma \frac{(Tj)^2}{nj} - \frac{(T)^2}{n}\right]$$

$$= \Sigma x^2 ij - \Sigma \frac{(Tj)2}{nj}$$

Coding Method

It is an extension of short cut method, does not change, if all the n item values are either multiplied or divided by common figure.

On this method big figures are reduced in magnitude by division or subtraction and and work is simplified without disturbance of 'F ratio.

Once figures are reduced in this method all the steps and shortcut method can beadopted.

Example :

The following table provides information about per hectare production of cotton of three varieties grown on four plots. Set up analysis of variance table and state, if the variety difference are significant.

Production' Table of Cotton

Production per hectare in Quantete

Plot	**Variety of Cotton**			Total of rotio
	H4	BT	Laxmi	
1	12	10	10	32
2	14	10	08	32
3	06	06	06	18
4	16	14	08	38
Total of coloumns	48	40	32	120

Solution

This problem can be sloved by both meytod and results will be same in case of appalication of each method.

1. Direct Method :

$$x_1 = \frac{12+14+6+16}{4} = 12$$

$$x_2 = \frac{10+10+6+14}{4} = 10$$

$$x_3 = \frac{10+8+6+8}{4} = 8$$

Mean of the Sample mean = $\bar{\bar{x}} = \frac{\bar{x}_1 + \bar{x}_2 + \bar{x}_3}{K}$

$$\frac{12+10+13}{3} = 10$$

SS between and SS within can be worked out

$$\text{SS between} = \bar{n}_1(\bar{x}_1 - \bar{\bar{x}})^2 + n_2(\bar{x}_2 - \bar{\bar{x}})^2 + n_3(x_3 - \bar{\bar{x}})^3$$

$$= 4\,(12-10)^2 + 4(10-10)^2 + 4\ \ (8-10)^2$$

$$= 10+0+16 = 32$$

SS Within

$$= \Sigma\,(xij - \bar{x}_1)^2 + \Sigma\,(xij - \bar{x}^2)^2 + \Sigma\,(xij - \bar{x}_3)^2 = 1.2.3.4.$$

$$= (12-12)^2 + \ (14-12)^2 + \ (6-12)^2 + (16-12)^2$$

$$+(10-10)^2 + (10-10)^2 + (6-10)^2 + (14-10)^2$$

$$+(10-8)^2 + (8-8)^2 + (6-8)^2 + (8-8)^2$$

$$\left[0+(2)^2+(-6)^2+(4)^2\right]\left[(0)^2+(0)^2+(4)^2+(4)^2\right](2)^2+(0)^2+(2)^2+(0)^2$$

$$= [4+36+16]+[16+16]+[4+4]$$

$$= 56+32+8$$

$$= 96$$

SS for total variance

$$\frac{16/2}{8}$$

$$= (12-10)^2 + (14-10)^2(6-10)^2 + (16+10)^2$$

$$+(10-10)^2 + (10-10)^2(6-10)^2 + (14+10)^2$$

$$+(10-10)^2 + (8-10)^2(6-10)^2 + (8-10)^2$$

$$= (2)^2 + (4)^2(4)^2 + (6)^2$$

$$= (0)^2 + (0)^2(4)^2 + (4)^2$$

$$= (0)^2 + (2)^2(4)^2 + (2)^2$$

So

$$= 4+16+16+36$$

$$= 16+16$$

$$= 4\times 16+4$$

So SS for total 1 28 = 32+96

On the basis of this problem Anova table can be set up as under

Sourcce of Variation	SS	df	MS	'F' ratio	5% limit From the 'F' Table
Between	16	(3-1)=2	$\frac{16/2}{8}$	8/5.33	F (2,9)
Sample	=1.5	=4.26	as for		
Within	48	(12-13)-9	$\frac{48/9}{5.33}$		table Sample value
Total	64	(12-1)=11			

The table sevels that 'F' ratio is 1.5 which is less than. 4.26 value calculate f 5f limit statistical table having (dF, 2 df, 9)

The interences drawn from the table is that suppor null hypotheses so we summarised that diffeence in cotton production due to varites is significant.

Short cut Method

The problem can also be clove through short cut method

1. Calculate of the individual value n item known as T

T in above care is = 120

andn = 12

Based on the same example calculate of 2 way Anova :

Step (I) T = 12 n= 12

$$\text{Correction Factor} = \frac{(T)^2}{n} = \frac{120 \times 120}{12} = 1200$$

Step (II)

Total SS

$$\begin{aligned} 144 + 100 + 100 &= 344 \\ 196 + 100 + 64 &= 360 \\ 36 + 36 + 36 &= 516 \end{aligned}$$

1,320 - 1,200 = 128

Step (III)

SS. between

$$\text{columns treatment} = \left[\left(\frac{48 \times 48}{4}\right) + \frac{(40 \times 40)}{4} + \frac{32 \times 32}{4}\right] - \left(\frac{120 \times 120}{12}\right)$$

$$= 576 + 400 + 256 - 1200$$

$$= 1232 - 1200 = 32$$

Step
SS between

coloumns treatment $= \left[\left(\frac{32\times 32}{3}\right)+\frac{(32\times 32)}{3}+\frac{18\times 18}{3}\right]-\left(\frac{120\times 120}{13}\right)$

$= (341.33 + 341.33 + 108 + 481.33) - 1200$

$= 1272 - 1200 = 72$

Step(V)
SS restdual = Tatal SS - (SS between column + SS wothu row)^R
or Error = 128- (32+72)
= 128-104
= 24

Example : 3

The following table gives information about three different drugs administered in different types of group of people.

Blood Pressure Reduction mercury (in mm)

Drugs

Group	A	B	C	Total row-wise
I	16 17	12 11	13 13	82
II	14 13	9 10	12 13	71
III	12 13	13 13	10 9	70
Total of columns	85	68	70	223

Find out and assess

(i) do the drugs act differently
(ii) different group people act differently
(iii) is interaction term significant
(iv) Significance level of 5'3F

Solution - step wise in v steps

Step I

T= 223 n=18

So correction factor $= \frac{223\times 223}{18}$

$= 2762.72$

Step IV

SS between rows

$$= \left[\frac{82\times 82}{6}+\frac{71\times 71}{6}+\frac{70\times 70}{6}\right]-\frac{(223)^2}{18}$$

$$= [1120.66+840.16+816.16]-2762.72$$

$$= 2776.98-2762.72 = 14.26$$

Step IV

SS between rows

$$= (16)^2+(17)^2+(14)^2+(13)^2+(12)^2+(13)^2$$

$$= (12)^2+(11)^2+(9)^2+(10)^2+(13)^2+(13)^2$$

$$= (13)^2+(13)^2+(12)^2+(12)^2+(10)^2+(9)^2 \quad -\frac{(223)^2}{18}$$

$$\begin{aligned}
&= 256+289+196+169+144+169 = 1223\\
&= 144+121+81+100+169+169 = 784\\
&= 169+169+144+144+100+81 = 807\\
&\qquad 2814
\end{aligned}$$

So 2814 - 2762.72

= **51.84**

Step V -

SS between sample

$$= (14-13.5)^2+(13-13.5)^2+(9-9.5)^2+(10+9.5)^2+(12\text{-}12.5)^2+(13-12.5)^2$$

$$= (12-12.5)^2+(13-12.5)^2+(13-13)^2+(13+13)^2+(10-9.5)^2+(9-9.5)^2$$

or $= (-.5)^2+(.5)^2+(.5)^2+(-.5)^2+0+0$

$$= (.5)^2+(-.5)^2+(-.5)^2+(.5)^2+(-.5)^2+(.5)^2$$

$$= (-.5)^2+(.5)^2+0+0+(.5)^2+(-.5)^2$$

$= 0.25\times 14 =$ **3.5**

Step VI -

S.S. for Interaction Variation

$$= 51.28-[28.76+14.26+3.5]$$

$$= 51.28-46.52$$

$$= 4.76$$

Anova Table

Variation source	S.S.	d.f.	M.S.	f ratio	5% limit
Between columns	28.76	(3-1) = 2	$= \frac{28.76/2}{14.385}$	$= \frac{14.385}{0.389}$	= f(2,9) 4.26
Between rows	14.26	(3-1)=2	$= \frac{14.2612}{7.13}$	$= \frac{7.13}{0.389}$	= f(2,9) 4.26
Interactor With	4.76	4	$\frac{4.7614}{1.19}$	$\frac{1.19}{0.389}$	f (4,9) = 3.63
Sample	3.50	18.9= 9	$\frac{3.0519}{0.389}$	-	
Eroor Total	51.28	(18-1)=17			

The above table indicate the fact that 'f ratio are significant at 5% lave] which revels that

(i) drugs act differently

(ii) different group people are affected differently

(iii) inter action term is significant

Interpretation of data

In the procedures of research methodology after collection and analysis of data research has to draw the inferences, and bases on these inferences, he will able to draw conclusion. This will provide him based to make recommendations.

Interpretation has two major aspects first to establish the continuity of the research by linking the results of study and secondly establishing the relationship with collected data.

Through interpretation the researcher can understand the real significance of his findings. Interpretation is the search for narrower meanings to his findings.

Interpretation enables the researcher to establish concepts, which will be foundation for further research to he conducted . Therefore it is an instrument through which the factors which are to be explained has been observed in research study.

Necessity of Interpretation

Interpretation is necessary for utility of research findings.

(a) Interpretation helps the researcher to understand the principle that work for his findings and through this, it helps to rationalized the research findings.

(b) Interpretation lead to the establishment of explanatory concepts and this beings base fir future research study. Through interpretation researcher can explain process, which has provided him deeper understanding relations, an factors operating in the study.

(c) By interpretation researcher can make other to understand importance of his findings.

(d) Interpretation is involved in the transition form exploratory to experimental research.

Techniques of Interpretation

Interpretation being a skilled work it requires skill on the part of the researcher. It is only though the application of various techniques which researcher will acquire necessary skill for interpretation.

The techniques of interpretation are given here through following steps involved in it.

Step I

Researcher must provide for relationship between to variables. This relationship may be of three types - symmetrical, reciprocal, asymmetrical.

The interpretation of data can be made on the basis of. these types of relationship interpretation,

at this stage is technique, as to how generalization should be done and concepts be formulated.

Step II

The compression between two or more series can easily he done by making use of percentages. These percentages can also describe the relationship. The following table shows the use of percentage. **For example - Sale two type of cars in major cities in Maharashtra**

Cities in	Car	P.C.	Car PE	PC
Maharashtra	A		B	
Mumbai	500	50	700	35
Pune	200	20	350	17.5
Nagpur	150	15	400	20
Aurangabad	80	08	300	15
Kolhapur	70	07	250	12.5
Total	1000	100	2000	100

It is very easy to understand sales of car A and B in various cities in Maharashtra on the basis of above table.

Step III

Average are also used for interpretation. There are three main forms of averages used for interpretation, such as– arithmetic mean, median and mode. These are several other measurers of central tendency. These three types of averages are commonly used, because interpretation through this very simple.

Step IV

After the percentages and averages dispersion is used for interpretation of data, dispersion refers to magnitude of the spread.

Measures of dispersion includes range, interquartile range, average deviation, standard deviation. With the help of this measures dispersion more scientific interpretation can be mode.

Step V

In order to understand the problem under study some extraneous information might have been
collected. One consideration should be given to this information while interpreting the study results.

Step VI

Sometime errors and omission are difficult to locate in such case, it is always advisable to take the help of person who will honestly point out such mistakes. This will not only helpful in correct interpretation but it will also enhance the research utility.

Step VII

All relevant factors affecting the study should be considered in order to avoid the false generalization. While interpreting, the result, it may be possible that the things which seems to be correct in the begging may not at all be correct.

Preventive measures for Interpretation

In spite of proper collection and analysis of data wrong interpretation will make our conclusions wrong.

Therefore it is necessary to have correct interpretation of data which is possible only when researcher has high degree of judgment and objectivity.

The following are the preventive measures in order to have appropriate interpretation.

1. Researcher must verify and satisfy himself regarding data being appropriate and adequate. It must reflect uniformity. Proper statistical methods should he used for analyzing data.
2. Precaution must be taken by the researcher about errors than can arise in results interpreting process.
3. Researcher must take up the work of interpretation as special task of analyzing.

   He must be careful enough regarding validity reliability of data and comparison of results.
4. Researcher is suppose to make not only sensitive observation but also find out to hidden factors, this will make job of interpretation on proper lines.
5. Research study in related to constant interaction between initial hypotheses empirical observation and theoretical conceptions. Researcher must give special consideration to all these factors while making interpretation.
6. Researcher must make sure the adequacy of data which should be large and unbiased, than only the result will represent the universe.
7. Understanding of definition, if results is in variations, it will lead to impropriates comparisons.

   There researcher should use the common understanding of the definitions so as to avoid improper comparisons.
8. Proper care should be taken in application of statistical tools. Inappropriate techniques and sample, will lead to wrong results. So in order to avoid error proper use of statistical methods should be made. This will be possible when research study includes large number of respondent.

QUESTIONS

1. (a) Explain the meaning of analysis of variance. Describe briefly the technique of analysis of variance for one-way and two-way classifications.

   (b) State the basic assumptions of the analysis of variance.

2. Below are given the yields per acre of wheat for six plots entering a crop competition, three of the plots being sown with wheat of vairety A and three with B.

Variety	Yields in fields per acre		
	1	2	3
A	30	32	22
B	20	18	16

Set up a table analysis of variance and calculate f. State whether the difference between the yields of two varieties is significant taking 7.71 as the table value of F at 5% level for $v_1 = 1$ and $v_2 = 4$.

3. The following table shows the lives in hours of randomly selected electronic lamps from four batches :

Batch	Lives in hours							
1	1600	1610	1650	1680	1700	1720	1800	
2	1580	1640	1640	1700	1750			
3	1450	1550	1600	1620	1640	1660	1740	1820
4	1510	1520	1530	1570	1600	1680		

Perform an analysis of variance of these data and show that a significance test does not reject their homogencity.

■■■

CHAPTER

9 CHI-SQUARE TEST

After going through this chapter, you will be able to know :

- Uses of Test
- Non - Paramatic Test
- Chi-Square Formula
- Expected Frequience
- Degree of Freedom (dt)
- Level of Significance
- Interence

The Chi-square test is one of the test of significance. It is written as χ^2 (pronounced as 'ki' square). It is also

called as test of association. It is used in relation to sampling analysis for comparing variance to a theoretical variance.

When we have nominal data in form of frequencies and when each and every observation is independent of all other observation in such cases this is applied.

This test can also be used to decide if two classifications are independent as a non - parametric test. All the researcher make use this test for the

a - goodness fit test

b - significance of assoication between two attributes.

c - significance of population variance.

When we used this test for the significance of population variances we workout value of χ^2 to test null hypotheses as under

$$\chi = \frac{\partial^2 s}{\partial^2 p}(n-1)$$

$\partial^2{}_s$ = sample variable

$\partial^2{}_p$ = population variable

(n – I) _) n = number of items in the sample i.e. degree of freedom.

Example :

Weight of 7 persons is given as below :

Persons	Weight
1	40
2	45
3	50
4	60
5	55
6	52
7	48

In this information we can say, variance of distribution of sample of 7 persons was drawn is equal to weight of 30 kg.

Test this at 5% of 1% level of significance.

Solution

Above information we will workout variance of sample data.

S.No.	X1	w/kg)	(x,-x)	(x,-x)²
1	40	-10	100	
2	45	-05	25	
3	50	+0	0	
4	60	+10	100	
5	55	+05	25	
6	52	+02	4	
7	48	-02	4	

$$n = 7 \quad 350 \qquad \Sigma(x_1 - \bar{x})^2 = 258$$

$$\bar{x} = \frac{\Sigma x1}{n} = \frac{350}{7} = 50$$

$$\partial^2_s = \sqrt{\frac{\Sigma(x_1 - \bar{x})^2}{n-1}} = \frac{\sqrt{258}}{7-1} = 43$$

$$H_0 = \partial^2 p = \partial^2 s$$

$$\lambda^2 = \frac{\partial^2 s}{\partial^2 p}(n-1) = \frac{43}{30}(7-1) = 8.6$$

Degree of freedom is(n – 1) = (7 – 1) = 6

At 5 Ye level of significance x2 = 12.592

1% level = 16.812

for 6 dF both the values

are greater than x2 = 8.6

So we accept null hypotheses and variance at both 5 and I pc level is significant. so sample of 30 kg is taken from the population.

Non - Paramatic Test

chi - square test is important non-paramatic text which can be conducted as
A- Chi - square Goodness Fit test -
B Chi - square test of association between two variables.

A. Chi- square Goodness Fit test -

This test is used where we can compare an observed frequency distribution to a theoretical frequency distribution.

When data consist of the observed frequency for a number of mutually exclusive categories is available and we want to know if they all occur equally frequently.

This concept can he very well illustrated by the various examples.

1. Suppose we want to find out relation between age of driver and probability of accident. In this case if we take data of 1000 accidents and age categories of driven as 21-30. 31-40, 41-50, 51-60, 61-70.

   If there is no relation between rate of accident and age of driver then there will be similar number of drivers in each category (i.e. - all categories should occur with equal frequency) If on the other hand younger driver are more lively to have accidents then there should be large number of accidents in this younger age categories and - low number of accidents in the older age categories.

2. Let us consider the fact that collage wanted to know which type of transport is used by student for coming to college. This questions will provide various alternatives as studies may use car, motor cycle. bus cycle.

   Take a sample of 100 students they may come by any of the about transport say 20 are using car, 60 motor cycle, 10 are coming by bus and 10 are using cycle.

   Each of the student would therefore fall into one of these categories.

   In this example we have one independent variable (type of transport) and (level of transport can be) Now if we want to know if there is preference in level of transport using car or bus or motorcycle.

   If in this case no preference is recorded it means " all categories occur with equal frequencies".

3. Suppose if we want to know certain college more popular with students for the purpose of admission. In this example independent variable is collage of study and level of this variable different college. M (Bombay). P (Pune). (A-(Aurnagabad) (Am) Amraoti and Nagpur (Ng.)

Our data consist of number of application for admission to each college. we will he comparing these observed frequencies to those that would occur, if all colleges were equally popular. Chi-square 'Goodness of Fit test' - Institute level.

We select 100 application of admission to these colleges and find that they are distributed across five colleges.

Management colleges at :

	Mumbai	Pune	Aurangabad	Amraoti	Nagpur		Total
Observed frequency	40	35	10	5	10	=	100

Are all colleges are equally popular! or some colleges attract more application than other.

As frequency shown above it appears that Mumbai and Pune colleges are more popular than other colleges.

We can examine this observation with application of Chi-square test on this data to find out if our observation deviates significantly from

We might expect to obtain if all colleges are equally popular with aplicants.

Chi-square formula :

$$\chi^2 = \Sigma = \frac{(O - E)^2}{E}$$

Step I Calculate the expected frequencies (Figure we expect if there were no college preference)

$$\textbf{(EF) Expected frequience} = \frac{\text{total number of ins tan ces}}{\text{number of categories}}$$

We have 5 categories (5 collages so

$$\text{E.F.} = \frac{100}{5} = 20$$

	Mumbai	Pune	Aurangahad	Amaravati	Nagpur	Total
O.F. (Observed Frequency)	40	35	10	5	10	100
Step I E.F. (Expected Frequency)	20	20	20	20	20	100
Step II substract OF - EF O-E	20	15	-10	-15	-10	
$(0\text{-}E)^2$ Step III $(OE)^2$	400	225	100	225	100	
Step IV E (20)	20	11.5	5	11.5	5	

Step 5 - Add all values of Step IV

$X^2 = 20 + 11.25 + 5 + 11.25 + 5 = 52.5$

This is obtained value it is single number discrepancy between U.E. and E.E.

To assess value represent real deposture in OE/EE it is to obtain various values Chisquare by chance.

1. Obtained value Chi-square - 52.5
2. Five categories so dF (degree of freedom), $(N-1) = 5-1 = 4$

3. Caculate dritical values Chi-square by statistical table

Value of 4 df at 0.5% = 9.49

at 0.1% = 13.28

0.01% = 18.46

If we compare obtained value 52.5 to this critical values it is much bigger the 'C' values. We conclude that 'OF' are significantly different than EF if all categories were equally popular. So we can say that all colleges are not equally popular with applicants.

(B) chi-Square test - Association between two variables

This method is used when have categorical data for two independent variable and if we want to workout association between them.

Problem 1

It we take same case for 'Institutional level admission' to management college as various places with two independent variables

Say 1 (χ) '9LA' applicants with independent variables.

(i) educational background - Commerence. Engineering

(ii) Choice of m'college - Mumbai, Pune, Aurangabad, Amaravati. Nagpur (Five levels)

Our observation frequencies can be arranged into following contengency table.

Management Colleges

Colleges	Mumbai	Pune	Aurangabad	Amaravati	Nagpur	Total
Commerce	3	1 0	12	5	40	70
Engineering	20	06	2	1	1	30
row total	23	16	14	6	41	100

The Chi-square formula is the same as explained earlier, only change is that expected frequencies

are calculated here.

$$\text{(EF) Expected frequence} = \frac{\text{total number of ins tan ces}}{\text{number of categories}}$$

Above table is reorganised with expected frequencies shown the brackets just below the obtained frequencies.

Management Colleges

Colleges	Mumbai	Pune	Aurangabad	Amaravati	Nagpur	Row total
Commerence	3 (16.1)	10 (11.2)	12 (9.8)	5 (4.2)	40 (28.7)	70
Engineering	20 (6.9)	6 (4.8)	2 (4.2)	I (1.8)	I (2.3)	30
Column total	23	16	14	6	41	100

Thus expected frequency of commerce students applying to mumbai college is 75 (row total) times 40 (coloum total) divided by 100 grand total.

$$80\text{ EF} = 80\text{ EF} = \frac{\text{rowt} \times \text{column t}}{100}$$

$$\frac{70 \times 23}{100}\ 16.1$$

$$\frac{30 \times 23}{100}\ 6.9$$

O–E (subract observed frequency from Expected F)

With this requirement now table will be set up as under.

With this requirement now table will be set up as under.

		Mumbai	Pune	Aurangahad	Amaravati	Nagpur
Commerce	O	3	10	12	5	40
	E	16.1	(112)	9.8	4.2	28.7
	O-E	13.1	12	2.2	08	11.3
Engineering	O	20	6	2	1	1
	E	6.9	4.8	4.2	1.8	2.3
	O.E.	13.1	12	2.2	0.8	11.3
Step III	$(O-E)^2$					
Com.		171.61	1.44	4.8	.64	127.69
Eng.		171.61	1.44	4.8	.64	127.69
Step IV	$\frac{(O-E)^2}{(OE)^2} \times E$					
Commerce	171.61/16.1	1.44/11.2	4.8/9.8	.64/42	127.69/28.7	
	10.659	123	.494	152	4.49	
Eng.	171.61/6.9	1.44/4.8	48/4.2	64/1.8	127.69/12.3	
	24871	.300	1.152	.356	10.381	

Problem 2

The marketing division of a company has drawn 1,000 random samples of five households each from different locations of a city to determine the number of households using its product. Here every unit of the sample consists of five households where the number of households using the product is denoted by x ($x = 0, 1, 2, 3, 4, 5$). The results are summarised in Table 5.20.

Table Number of Households Using the Company's Product

Number of Households Using the Product (x)	Number of Samples (Observed Frequency : f_0)
0	334
1	373
2	230
3	55
4	7
5	1
Total	1,000

In Table 5.20, the number 334 indicates that 334 households out of a sample of 1,000 households of size 5 each do not use the company's product. Similarly, the other observed frequencies may be interpreted. Is it correct to conclude that the data in Table follows a binomial distribution ?

As mentioned in Chapter............, the two parameters of the binomial distribution are the sample size (n) and the probability of the occurrence of an event (p), and the mean is np. Here n = 5 and p is the probability that a sample of five households will buy the product. The mean of the distribution in Table is.

$$[(0)(334) + (1)(373) + (2)(230) + (3)(55) + (4)(7) + (5)(1)] / 1{,}000 = 1.031.$$

p is estimated by equating np to 1.031. So $p = 1.031/5 = 0.21$.

From (22) in Chapter , the probability that x out of n families will buy the company's product is given by the binomial distribution

$$b(x; n, p) = [n!\ x!\ (n - x)!]\ p^x (1 - p)^{n - x}$$

Here the null and alternative hypotheses are H_0 : (observed distribution comes from a binomial distribution with parameters $n = 5$ and $p = 0.21$) and H_1 : (observed distribution does not come from a binomial distribution with parameters $n = 5$ and $p = 0.21$) respectively.

Calculations for Chi Square

Number of Households Using the Product *(x)*	Observed Frequency *(fo)*	Binomial Probability	Expected Frequency *(fe)*	f_0^2 / fe
0	334	0.3077	308	362.19
1	373	0.4090	409	340.17
2	230	0.2174	217	243.78
3	55	0.0578	58	52.16
4	7	0.0077	8	8.00
5	1	0.004	0	–
Total	1,000	10.000	1,000	1,006.30

From (2)

$$\chi^2 = \Sigma f_0^2 / fe - N = 1{,}006.30 - 1000.00 = 6.30$$

(Degrees of freedom = 5 - 1 - 1 = 3)

The critical values of χ^2 at 5% and 1% levels of significance and 3 degrees of freedom are $\chi^2_{0.05,\ 3} = 7.815$ and $\chi^2_{0.01,\ 3} = 11.345$.

Decision : Since the calculated value of χ^2 is less than its critical values, the null hypothesis H_0 : (observed distribution comes from abinomial distribution with parameters n = 5 and p = 0.21) is accepted.

Conclusion : It may be concluded that the observed distribution follows a binomial distribution with parameters n = 5 and f = 0.21.

Note 1 : The last two classes in Table have been merged to ensure that the expected frequency is not less than 5.

Note 2 : The degrees of freedom of χ^2 is 3 because the parameter p of the two parameters n and p of the binomial distribution has been estimated from the sample data.

Problem 3

The sales manager of a showroom feels that the daily demand of a product follows uniform distribution. The observed frequencies of demand values are summarized in Table 1

Table 1 Observed Frequencies of Daily Demand

Demand (units), (X_i)	Observed frequency, (O_i)	Demand (units), (X_i)	Observed frequency (X_i)
30	13	35	9
31	10	36	12
32	7	37	10
33	10	38	14
34	6	39	9

Check whether the given data follow uniform distribution at a significance level of 0.05.

Solution : The combination of hypotheses for this problem is :

H_0 : The given data follow uniform distribution

H_1 : The given data do not follow uniform distribution.

We also have

The total of the observed frequencies = 100

Number of demand values = 10

Expected frequency for each Xi = 100/10 = 10

The details of the calculations of the chi-square value are summarized in Table 2.

Serial number	Demand value, X_i	Observed frequency, O_i	Expected frequency e_i	$\frac{(O_i - e_i)^2}{e_i}$
1	30	13	40	0.9
2	31	10	10	0.0
3	32	7	10	0.9
4	33	10	10	0.0
5	34	6	10	1.6
6	35	9	10	0.1
7	36	12	10	0.4
8	37	10	10	0.0
9	38	14	10	1.6
10	39	9	10	0.1

The observed chi-square statistic = $\sum_{i=1}^{10} \frac{(oi - ei)^2}{ei} = 5.6$

The table chi-square value with 9 degrees of freedom at the given significance level of 0.05 placed at the right tail of the chi-square distribution is 16.919. This is shown in Fig.

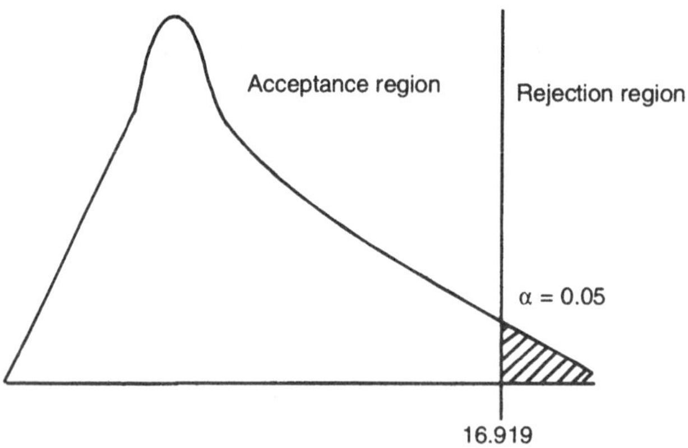

Fig. Chi-square distribution for Example

Since the calculated chi-square value (5.6) is less than the table chi-square value (16.919), it falls in the acceptance region. Hence, accept H_0.

Thus the given data follow uniform distribution.

Problem 4

The manager of a branch bank feels that the arrival rate of customers in his branch bank follows Poisson distribution. The observed frequencies of arrival rates (number of arrivals in 15 minues intervals) are as summarized in Table.

Table Observed Frequencies of Arrival Rates

Serial number,	Arrival rate,	Observed frequency,
i	Xi	Oi
1	0	2
2	1	4
3	2	8
4	3	14
5	4	7
6	5	5
7	6	4
8	7	3
9	8	2
10	9	1

Check whether the given data follow Poisson distribution at a significance level of 0.05.
Solution The combination of hypotheses for this problem is :

H_0 : The given data follow uniform distribution

H_1 : The given data do not follow uniform distribution.

We also have

The total of the observed frequencies = 50

Number of values for the arrival rate = **10**

Then

$$\text{Mean arrival rate, } \delta = \frac{\sum_{i=1}^{10} XiOi}{\sum_{i=1}^{10} Oi} = 3.7 \text{ arrivals per 15 } min.$$

The probability distribution for the Poisson distribution is

$$P(X) = \frac{\delta^x e^{-\delta}}{X!}, \quad X = 0, 1, 2, 3, \ldots.$$

The calculation of the chi-square value is summarized in Table

Table Summary of Calculations of Chi-Square Value

Serial number	Demand value, Xi	Observed frequency, Oi	Expected probability P(Xi)	Expected frequency ei	$\frac{(Oi - ei)^2}{ei}$
1	0	2	0.0247235	1.24	0.466
2	1	4	0.0914770	4.57	0.071
3	2	8	0.1692326	8.46	0.025
4	3	14	0.2087202	10.44	1.214
5	4	7	0.1930661	9.65	0.728
6	5	5	0.1428689	7.14	0.641
7	6	4	0.0881025	4.41	0.038
8	7	3	0.0465685	2.33	0.193
9	8	2	0.0215379	1.08	0.784
10	9	1	0.0088545	0.44	0.713

The observed chi-square statistic = $\sum_{i=1}^{10} \frac{(oi - ei)^2}{ei} = 4.873$

The table chi-square value with 9 degrees of freedom at the given significance level of 0.05 placed at the right tail of the chi-square distribution is 16.919. This is shown in fig.

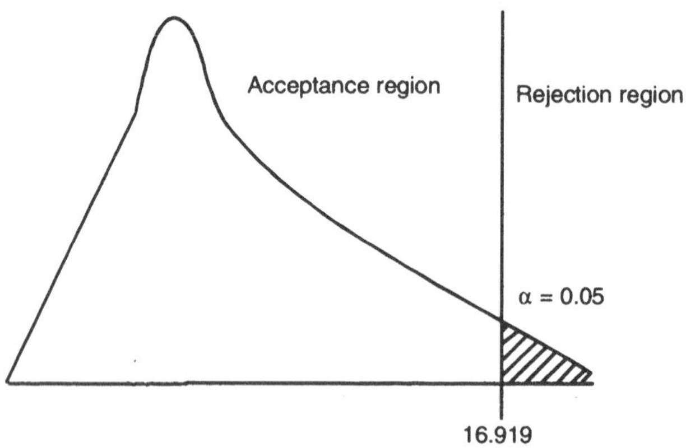

Chi-square distribution for Example

Since the calculated chi-square value (4.873) is less than the table chi-square value (16.919), it falls in the acceptance region. Hence, accept H_0.

Thus the given data follow Poisson distribution.

Problem 5

Suppose consumer preferences and place of residence have been cross classified the following data is given about three different product FMCG.

Table - 'A'

Relationship between consumers preference and place of resident regarding different product of fast moving consumer goods.

Place of Resident	Consumer Preference Product			Total
	A	B	C	
Urban	76	40	24	1240
Rural	20	52	12	84
Total	96	92	36	224

Based on above table it we take urban preference to FMCG. They are

A	B	C
$\frac{76}{96}$	$\frac{40}{92}$	$\frac{24}{36}$
= .79	= .43	= 66

if we want to know wether differences are significant so it null hypotheses is true than we can calculate as

$$\frac{76}{96}+\frac{40}{92}+\frac{24}{36}=\frac{140}{224}=0.625$$

Table A can be modified for expected frequences we expect

- 93 $\times$ .0.625 = 60 have choice for A.

- 92 $\times$ 0.625 = 57.5 have choice for B.

- 36 $\times$ 0.625 = 22.5 have choice for C

act of 140 urban population.

where as in case of rural population.

96.60 = 36 have choice for A.

92 - 57.5 = 34.5 have choice for B.

36 - 22.5 = 13.5 have choice for C.

This can be shown in Table B

Place of Resident	Consumer Preference Product			Total
	A	B	C	
Urban	76	40	24	140
	(60)	(57.5)	(22.5)	
Rural	20	52	12	84
	(36)	(34.5)	(13.5)	
	96	92	36	224

In order to test tenability of null hypothese comparison of observed and expected frequences is done with this formula

$$\chi^2 = \Sigma \frac{(O-E)^2}{E}$$

O = Observed frequences
E = Expected frequences

Table - 'C'
Calculation of X^2

P	R	O	E	O-E	$(O-E)^2$	$\frac{(Q-E)^2}{E}$
A	U	76	60	16	256	4.2666
	R	20	36	-16	256	7.1111
B	U	40	57.5	-17.5	306.25	5.3260
	R	52	34.5	-17.5	306.25	8.8768
C	U	24	22.5	1.5	2.25	0.1
	R	12	13.5	-1.5	2.25	0.1666
Total	224	224				$X^2 = 25.7871$

Degree of Freedom (df)

'df'–degree of freedom means the number of independent constraints imposed.

'df' can be calculated by this formula

df = (c–1) (r–1)

df–degree of freedom

C = number of columns

r = number of rows

Dn table–column = 3 row = 2

dt) (c–I) (r–l)

= (3–1) (2–1)

= 2 × 1 = 2

Level of Significance

che'-square test is used to examine whether the difference between observed and expected

frequencies is due to sampling fluctuations and such significant : Wether differences is due to some other reason, and such significant.

In the same example If we and want to calculate df and sig. it can be explained as under a Level significance ay 5% df - 'df'- Calculate as already expanded in Table 'A'

df–(E–1). (r–1) = (3.2)(2–I)=2×1 = 2

As per the statistical table level at sf at df²

0.05 = 5.99

Inferences

Since X= = 25.7871

level of significant 5.99 at sf with df²

Hence

I Null hypotheses differences among rural and urban population with respect to consumer preferences is effected.

-We therefore conclude that there is a association between place of residence (rural) and urban) and consumer preferences.

Problem 6

Table shows a random sample of people at different levels of education and their monthly incomes :

People in Different Categories

Level of Education	Monthly Income (Rs.)				Total
	< 5000	5000-6999	7000-8999	> 8999	
Undergraduate	55	65	40	30	190
Graduate	42	71	30	40	183
Postgraduate	25	38	43	21	127
Total	122	174	113	91	500

Test whether there is an association between the level of education and the monthly income.

Here the null and alternative hypotheses are H_0: (there is no association between level of education and monthly income) and H_1 : (there is association between level of education and monthly income) respectively. The expected frequencies are shown in Table.

Expected Number of People at Different Categories

Level of Education	Monthly Income (Rs.)				Total
	< 5000	5000-6999	7000-8999	> 8999	
Undergraduate	46	66	43	35	190
Graduate	45	64	41	33	183
Postgraduate	31	44	29	23	127
Total	122	174	113	91	500

Values of f_0^2 / fe are computed from Tables 5.14 and 5.15. These are shown in Table

Table Values of f_0^2 / fe

Level of Education	Monthly Income (Rs.)				Total
	< 5000	5000-6999	7000-8999	> 8999	
Undergraduate	65.76	64.02	37.21	25.71	192.70
Graduate	39.20	78.77	21.95	48.48	188.40
Postgraduate	20.16	32.82	63.76	19.17	135.91
Total					517.01

From (2)

$$\chi^2 = \Sigma f_0^2 / fe - N = 517.01 - 500.00 = 17.01$$

(Degrees of freedom = 6)

The critical values of χ^2 at 5% and 1% levels of significance and 6 degrees of freedom are $\chi^2_{0.05,6} = 12.592$ and $\chi^2_{0.01,6} = 16.812$.

Decision : Since the calculated value of χ^2 is greater than its critical values, the null hypothesis H_0 : (there is no assoication between the level of education and the monthly income) is rejected.

Conclusion : There exists a statistically significant association between the level of education and monthly income.

Problem 7

Table shows a random sample of 500 managers classified on the basis of years of job experience and levels held by them.

Table Classification of Managers by Years of Job Experience and Levels

Years of Job Experience	Levels				Total
	Top	Senior	Middle	Junior	
Less than 5 years	7	12	38	162	219
5 - 10 years	15	22	19	88	144
More than 10 years	35	48	20	34	137
Total	57	82	77	284	500

Does there exist an association between years of job experience and levels held by the managers?

Here the null and alternative hypotheses are H_0 : (there is no association between years of job experience and levels held by the managers) and H_1 : (there is association between years of job experience and levels held by the managers) respectively. The expected frequencies are shown in Table

Table Expected Frequency of Managers by Years of Job Experience and Levels

Years of Job Experience	Levels				Total
	Top	Senior	Middle	Junior	
Less than 5 years	25	36	34	124	219
5 - 10 years	16	24	22	82	144
More than 10 years	16	22	21	78	137
Total	57	82	77	284	500

Values of f_0^2 / fe are computed from Tables and these are shown in Table

Table Values of f_0^2 / fe

Years of Job Experience	Levels				Total
	Top	Senior	Middle	Junior	
Less than 5 years	1.96	4.00	42.47	211.65	260.08
5 - 10 years	14.06	20.17	16.41	94.44	145.08
More than 10 years	76.56	104.73	19.05	14.82	215.16
Total					620.32

From (2)

$$\chi^2 = \Sigma f_0^2 / fe - N = 620.32 - 500.00 = 120.32$$

(Degrees of freedom = 6)

The critical values of χ^2 at 5% and 1% levels of significance and 6 degrees of freedom

are $\chi^2_{0.005,\,6} = 12.592$ and $\chi^2_{0.01,\,6} = 16.812$

Decision : Since the calculated value of χ^2 is greater than its critical values, the null hypothesis H_0 : (there is no association between years of job experience and levels held by the managers) is rejected.

Conclusion : There exists a statistically significant association between years of job experience and levels held by the managers.

QUESTIONS

1. What is Chi-square test? Explain its significance in statistical analysis.
2. Write short notes on the following
   1. Conditions for applying Chi-square test
   2. Additive property of Chi-Square
   3. Chi-Square as a test of 'goodness of fit?

CHAPTER

10

RESEARCH REPORT AND PRESENTATION

After going through this chapter, you will be able to know :

- Report drafting
- Types of Reports
- Techniques
- Methods of Research Report writting
- Reference to Footnotes
- Table and Charts
- Bibliography and Index
- Diagrammatic Presentation
- Types of Diagram
- One-dimensional diagrams
- Two-dimensional diagrams
- Three-dimensional diagrams
- Pictograms
- Cartograms

The research report is the final stage to be completed in the process of the research. It is an important part of the research study. In research report, researcher will have to make complete account of the research activity conducted.

In research report the problem of research methodology used and conclusion drawn will be described. Writing of research report is different type of work in the research process. It requires to have different skill to complete the task, therefore due care should he taken in this stage of report writing.

Report drafting

Research report is an outcome of accurate work conducted during research process. The following steps are involved in drafting of the research report

(i) Statement of problem and its analysis.

(ii) Outline of research work.

(iii) Rough draft of the research work

(iv) Redrafting

(v) Bibliography

(vi) Final draft of the work.

(i) Statement of problem

The process of analyzing the subject matter is the first step in drafting of report. The subject matter is developed in statement of problem by two ways, one it should he developed logically and second chronologically. Logical analysis of subject matter from simple to complex one. It is based on associations between one and other method of analysis. Chronological development of subject matter is based on arrangement in occurrence.

(ii) Outline of the research work

It means the frame work within which report is constructed or written in particular order. This provide based for all those important points which are to be *highlighted* in the drafting of a report.

The outcomes of research works are to be conveyed to end-users or recorded for future use. The research efforts will be useful to the end-users only when they are documented in the form of reports. So, a standardized format for the report will help researchers present their contributions and findings more systematically.

The different items of a research report are presented in following format.

1. Cover page
2. Introductory pages
   - Vita
   - Acknowledgement
   - Abstract
   - Table of contents
   - List of tables
   - List of figures
   - List of variables
3. Text
   - Chapters
   - main sections
   - subsections
   - Footnotes
   - Conclusions
4. Bibliography
5. Appendices

(iii) Rough draft

The next stage now in the drafting of report is rough draft which is final outline of the work. In this stage researcher prepare the draft regarding conditions and circumstances of the research study.

In rough draft researcher will explain how material was called by the for the purpose of the study and various methods he has adopted for analyzing the martial collected by him. What are its observations in the study based on which conclusions are drawn and reconditions are made.

Table of Contents of Algorithmic Research Report

Table of Contents

Exhibit of Cover Page and Table of Contents of Survey-Based Research

Developement of Agro-Industries in Nagpur District since 1960

A Thesis submitted to Nagpur
University for the award of
Degree of Doctor of Philosophy
in the faculty of Commerce

Under the Supervision of
Prof. M. R. Tokhi
Nagpur.

By
V. K. Upagade
Nagpur

Exhibit of Cover Page and Table of Contents of Survey-Based Research

Financial Structure of Nagpur Municipal Corporation

A Thesis submitted to R.T.M., Nagpur
University for the award of
Degree of Doctor of Philosophy
in the faculty of Commerce

Under the Supervision of
Dr. V. K. Upagade
Nagpur.

By
A. K. Shende
Nagpur

(iv) Redrafting of the report

Once the rough draft of the research report is prepared it requires to be updated. So continuous revision is required in the rough draft of the report.

Redrafting of the report will help the researcher to examine the accuracy and quality of the research work. Sometimes redrafting requires more time than of rough draft. Researcher at this stage should also examine the consistency in the rough draft along with grammatical composition of the work.

(v) Bibliography

The reference in the research work is one of the important factor. The material collected for the purpose of the study might have been drawn from various sources. So the reference to these sources are essential in research study.

The list of published and unpublished sources in from of books papers are the main contents of the bibliography. This should be arranged alphabetically. First part should be for the references of book and second part reference should be made to papers magazines Reference should also be made if some unpublished record is used in the research study.

There is no specific system in which references are to be made in respect of books or magazines, journals, news papers adopted for the work. The normal practice adopted for reference is as under -

In cases of references made from books

1. Author of the book - name.
2. Title of the hook.
3. Publisher of the hook.
4. Place and year of publication
5. Volumes

Example

Upagade V.K. 'Problems of Income Tax' Anuradha Prakashan, Nagpur, 2007, 3rd Edition. For Journals, newspapers and magazines

1. Author's name
2. Title of article
3. Name of periodical
4. Publishers name
5. Year - month - date of publications
6. Place of publication
7. Page number

Example

Deokota SR. Socio-Economic development in Nepal, South-asia Economic Journal, Vol and Not, July. Dec'07. Pg 285

Example given above are only samples whatever method is adopted to make references should be consistent.

(vi) Final draft of the research work

The main thrust in any of report writing is language which should be very simple and self explanatory. Avoid the used of complicated terms and vague expression.

Wherever illustration is given it should not be imaginary. This must be based on fact. The final draft of the report must be effective in communicating the findings of the research work.

In research report the problem which is analyze must contribute to the solution to the problem and findings in report should enrich the knowledge and must he useful to all.

Types of Reports

Research reports differs on account of its length, form and situation of research. The report may be short or long.

(A) Short report consist of brief statement. A research study and examination of research problem in its all details. This will include the statement of main points along with conclusions and suggestions.

If the problem is well defined in the study the short report are appropriate in such cases. The studies with limited scope have simple methods can easily cover in short report.

(B) Long report may be in form of technical or popular report. **Technical report** is prepared for record keeping or for spreading in public. **Popular report** is prepared, in case if research results are based for implications of policy. Both the reports are explained as under -

(i) Technical Report

The technical report consist of the three main factors i.e. methods, assumptions and presentation.

The main findings including their limitation and supporting data should be given in the beginning.

Objectives and general description of the formation of problem and hypotheses, collection and analysis of data to be explained.

Detailed information regarding sample design, sample selection should be maintained.

The source of data, primary or secondary, as required or the study their limitations, suitability should be assessed and properly described.

The analysis of data and presentation of findings along with supporting chart table should be narrated.

Finally the conclusion drawn from the results be explained. Bibliography appendices and index should be prepared at the end.

(ii) Popular Report

The popular report consist of the simplification and attractiveness. Simple and clear writing, minimum use of mathematical and technical expression are the main aspects of simplicity. Another characteristic is attractive layout of the report.

The popular report should be compiled as under–

The emphasis should be given on the main findings and its implications. It should also suggest the action to be taken on the basis of the findings.

The objectives of the study should maintain how problem arises and presented in technical study. Which methods are used for data collection now analyze, should be explained.

Lastly in the main part of the report study should be presented in simple and clear terms by making use of charts, diagrams as required. Appendices includes the information regarding forms used in the study at the end.

There can be various forms of completing technical or popular report. The only important fact that nature of the study. methods use of collection and analyzing data and summary of conclusion and suggestion should be properly explained.

Contents of the Research Report

The research report should start with problem selected for research study. The report should consist of nature, scope and limitations of the proposed study. This will help to understand the importance of the study conducted.

If the investigation is planed with a view to make contribution in the theory, than theory within which study is organized should be explained.

In order to have proper presentation and easy understandability research report format consist of the following aspects.

1. Preliminary Text
2. Main text
3. Reference material

1. Preliminary Text

The composition of preliminary text include the title page. preface or foreword and acknowledgements. It also includes table of contents, list of charts, graphs figures used in study.

2. Main Text

It provides the complete outline of the research work. It will include introduction, main body of the report, findings and conclusions, summary of the results.

3. Reference Material

This will include the bibliography appendixes and indexes. This is also know as end matter. All these three main aspects of the contests of the research report are described here separately.

1. Preliminary Text

As it has been already explained in the beginning that the preliminary test in related to title and preface of the research work. Detailed information in this respect is given as under.

A. Title of the research work will include name of the researcher the organization of researcher place and year of the research.

B. Preface or forward relates to the acknowledgements in selection to the study conducted. This will also include object nature and scope of the study.

It is necessary to make the references about the persons and organization who have extend their co-operation in compiling of the work. The assistance or the guidance provided to by them in making study more useful.

Special mention should be made if financial assistance is provided by any of the organization. The appreciation of the co-operation extended in any from should be properly recognized in preface note.

C. Table of contents - The main object of table of contents is to provide overview of the material included in the research study. This is basically deals with scheme of chapteriasation with brief information included in each of the chapter.

The entire study is required to divided in various chapters. First being introduction and last is regarding observations conclusions are developed according of the need of the study.

Though table, charts and graphs are used in various chapters for presenting information in tabulated form but reference to all the charts in various chapter should be mode available at one place. So the list of the tables charts and graphs is prepared so the reference will help to anybody who is interested to find out information he can easily locale the information in the research study report.

2. Main Text of the research report

This is an important aspect of the research study it provides the outline of the research with more details of the study through the scheme of the chapterisation. It will consist of the following sections.

A. Introduction:

The main purpose of the introduction is to make clear statement of the objectives of the research. It should also include information about other relevant research as to how present study can he seen in the light of the content.

If proper introduction is not made it will make our research study aimless. In order to have clarity in research proper explanation of definitions and concepts used in research should be made. In conducting the methods adopted should also be explained.

This chapter contains the following information.

1. The statement of problem.
2. The scope of the study.
3. The need and importance of the study.
4. The historical background of the research problem.
5. The sources of data collected methods use for analyzing the data.
6. The various limitations under which research study conducted, must be narrated.
7. Questionnaire used for collection of data's what type of question were asked should be given in brief
8. The hypotheses of the study should be stated in this part.

B. Main part of the report

The main part of report consist of various chapters. The findings are incorporated in this part. Every statement made at this stage in relation to findings of the research study should be properly supported with the evidences and facts. If the findings are extensive than they should be presented in summarized form.

In this part of the report the following information should be included.

1. Detailed findings along with tables and charts.
2. Results in logical sequence.
3. presentation of relevant data.
4. Interpretation of data should include.
   a. Statement of inferences.
   b. Conditions which limits the inferences.
   c. Unanswered questions.

C. Conclusion of the study

The research study report should be concluded with brief account of the activity conducted: methodology used and finding and conclusions are drawn. This should also highlight potential of the further research in the context of the study. The necessary instructions in this respect to future action should in incorporated in this stage.

3. Reference Material

The reference material consist of the bibliography. appendix and index At the end of the research studs report all this information should be provided.

A. Bibliography

It is min part of the reference material. In compilation of the bibliography information regarding primary and secondary sources text articles and journals can be classified accordingly.

There in no division in the classified structure of the bibliography regarding exhibits, but comprehensive listing of the sources should have logical arrangement in the proper alphabetical order in composition of the bibliography.

B. Appendix

This includes copy of questionnaire and interview schedule which are used in collecting the primary data :

This will also include tables as supporting evidence, test that have been constricted for the research study.

If some documents and original data is used, in order draw findings are also tested out in appendix

C. Index

If the research report or the theses is published in form of a book or in any other form like. bulletin the in such case, index is necessary. This act as guide for him regarding the contents of the research study.

Methods of Research Report Writing

The basic aspects related to techniques of writing report are clarity and accuracy. The criteria of the standardized format should be decided on the basis of the material collected. The following facts deserve due consideration while writing research report.

1. The paper used for writing of a report should have standard size and design. It must have proper margin top and the paper should be neat. The information on the paper may be in single or double space either on one side or on both the sides of the page.
2. The various steps in report writing should be strictly followed and layout of the report should be thought and decided and accordingly adopted.
3. If quotations are to be used for explaining the text than it should he given proper treatment. It should be placed in quotation marks.
4. The systematic presentation of report requires by employing heading, sub-heading small paragraphs. underline and bold lettering as per the requirement of the subject matter.
5. In writing stage of the report the facts should be given due considerations.
   a. Effective communication through report between reader and writer of the report.
   b. Limitation under which report is written.
   c. Utility of the report.
   d. Report writing should considered the above facts and make report interesting to read. In order to have effectively in the report. If quantitative information is collected then it should be presented in tables and graphs. The graphic presentation is more attractive and easy for interpretation.

6. References to Footnotes

In research study report if footnotes are used with an object of cross reference, authority of source and explanation to view expressed. Footnotes are normally placed at the page bottom.

The footnotes are numbered consecutively in each page separately. this should correlate to the reference in text.

7. Table and Charts

Statistical information in presented in table charts and graphs. This presentation should he complete and it should not require further explanation. Such presentation should be very clear.

8. Bibliography and Index

The preparation of bibliography and index has already explained earlier; according they should be prepared. There should be uniformity in the reference and attached to the research report .Preventive measures for writing of good research report-

There are no specific rules for writing of research report poor presentation of report will not able to impress the reader inspite of its good quality. As such researcher must considered following while writing of a report of research study.

1. The research report should be long enough for proper subject coverage at the same time it should be short enough to maintain the interest of the reader.
2. Use of technical terms should be avoided. Report should be as simple as possible.
3. Quantitative data should be presented in table charts and graphs to make presentation attractive.
4. Research report have appropriate layout according to the objectives of the study.
5. The composition of language should name effective communication. Efforts should be made to avoid grammatical errors.
6. Report must contain logical analysis of the subject matter.
7. Research study is carried out with a intention to find out the solution to the research problem. Research report should behave origionality of the work.
8. Research report must indicate the implications of the policy in relation to the problem under study.
9. Bibliography, appendices and index are the essential part of the report.
10. physical appearance of the report should be good to attract the reader.
11. Report should remind readers actual operation performed. Reader must have an opportunity to know the answer to question which surfaced as result of the study.
12. Report should go through continuous examination at various stage. This requires up dating of information to have utility of the reports.

As such writing of report a skill which can be developed by experience.

Problems in Writing a Report

Report writing is not an easy task. The main task before the writer is to communicate the subject-matter but the communication has to be done in such a manner that it can be followed by the reader. Apart from it, the generalization should be properly worded and communicated. The problems that are faced by the writer of the report may be studied under the following heads :

The problems of communication

The problem of communication is the basic problem of report-writing. The finding of the study or research are the technical knowledge or a technical matter. It is not easy for laymen to follow them. The investigator, while writing the report has to keep in mind the fact that the report is not meant only for technical hands or those possessing technical knowledge. He has, therefore, to take precaution in regard to the following :

(1) Technical terms should be properly explained : Technical terms are bound to be used in a report of study of social problem. Now the technical terms have been standardized and so there is not much difficulty in their use. But the investigator while writing the report, has to use these technical terms in a manner that they while communicating the findings, are also clear to the reader even if he is a layman.

In research, all the technical terms have not been fully standardized and this present's a problem in the writing of the report. The investigator has to be cautious while he uses these technical words.

(2) Neither too simple nor too difficult expression : If the investigator makes the report too simple, there is likelihood of its not carrying the required impression. On the other hand, if it becomes too terse or too difficult, it shall not be followed by many persons who are not specialist of the subject and yet interested in acquiring knowledge about the subject. It means that the investigator should follow the golden middle path while maintaining the technical character, he should try to write it and present it in an easy-to-follow manner. No doubt for reading a research report, the ordinary reader shall have to rise higher, but the researchers should also come down from their peak of intellectual specialization.

(3) Level of knowledge and subject - matter - White writing the report, the report, the investigator has to keep in mind the level of the knowledge of the persons for whom it is meant. All the reports are not meant for research students or scholars. They are also meant for general public. The investigator has to link the subject-matter with the level of the knowledge of the persons for whom the report is meant.

(4) Language and drafting - The problems of language and drafting are very important for a report. The language has to be such that it is neither too colloquial nor too scholarly. The technical words should be used but their meaning should also be explained. The language should be free from magnitude. It should also not be incomplete or written in a flowery language that relegates the facts to the background. The remarks or the conclusion should be written correctly and clearly.

Drafting of the report requires not only grasp over the subject-matter but also command over the language. Apart from it, the investigator has to keep his own feelings and emotions under control while he is writing the report. The report should not look like a statement about the facts collected but it should make a pleasant reading.

The problem of objectivity

In research, objectivity is a guiding factor. The presentation of the data collected should be very objective. It means that the report should be free from prejudices and bias and should have no place for exaggeration. When the report is written about the research in the field of natural sciences, the subject-matter helps the objective presentation of the data collected. But in research, the researcher or the investigator has to take extra precautions for maintaining objectivity.

The problems of expression of bitter truths and unpleasant facts

While the investigator is collecting data, he comes across all sorts of facts. Some of these facts are quite bitter, not only for certain group of people but also for the investigator himself. The investigator has to present them in his report. Unless the conclusions that are based on the entire data collected are given in the report, the report shall be lop-sided. In presentation of bitter facts and unpleasant truths, the investigators has to do tight-ropewalking. It means that he has to present these facts in a manner that they neither injure feelings of a particular group nor do they harm any group.

Apart from it, sometimes the investigator is faced with the problems of studying certain facts that are likely to create great commotion. If these facts do not affect materially the findings of the research, he can safely hold them back. Similarly, the investigator has to be very cautious in regard to half-truths or facts which cannot be proved but are generally believed as true. If the investigator keeps back these half-truths, he does not provide proper guidelines to future researchers. There again he has to be very cautious.

Careful Consideration of Research Report

Research report is a statement for communicating research findings. Therefore research report should give careful consideration to the following important aspects.

1. The research report should generate sufficient interest of the reader. The report should neither be very lengthy nor very short. At the same time it should cover all the points keeping in view the interest of the readers.
2. As for as possible use of technical terminology should be avoided.
3. The charts, graphs and tables should be used as per the requirement of the presentation in order to make explanation easy.
4. The layout of the report should be properly planned.
5. Efforts should be made to avoid the grammatical mistake part of the report and they should be recorded as per the standard norms.
6. There should be logical analysis in the report.
7. Appendices, Bibliography and Indexing are the integral part of the report and they should be recorded as per the standard norms.
8. The report must be attractive in its appearance.
9. Research report must state the objectives and nature of the research problem under study.

Though all these points are the guidelines of the report writing, it is the normal practice of the researcher that he will able to draft good report by his experience only.

Report Presentation

After the preparation of the report, particularly research report it's presentation is also equally important aspect of the research activities. General write-up of the report is sufficient enough to understand the contains of the research study conducted.

There is no hard and fast rules as to how report is to be presented, but if it is presented through graph and diagram it becomes easy to understand. So normally every researcher will design research report with suitable diagrams and graphs.

Therefore it is necessary to know how to distinguish between diagrams and graphs. The graph is constructed to understand the mathematical relationship between two variables, where as diagrams are constructed to understand the comparisons.

The presentation of diagrams are made by bars, squares, circles, cubes and rectangles and graphs presented by points of line like dots and dashes. The graphs are used for the study of time series and frequency and diagrams can only depict categorical and geographical data.

The graphs are easy to construct than of diagrams. In this chapter we will first find out various types of diagrams which can be used for proper presentation of latter on we will work out different methods for graphic presentation.

DIAGRAMMATIC PRESENTATION

General Rules for Constructing Diagrams

1. Neatness, As already pointed out, diagrams are visual aids for presentation of statistical data and are more appealing and fascinating to the eye and leave a lasting impression on the mind. It is, therefore, imperative that they are made very neat, clean and attractive by proper size and lettering; and the use of appropriate devices like different colours, different shades (light and dark), dots, dashes, dotted lines, broken lines, dots and dash line, etc., for filling the in between spaceof the bars rectangles, circules, etc., and their components. Some of the commonly used devices are given below

2. Title and Footnotes

As in the case of a good statistical table, each diagram should be given a suitable title to indicate the subject-matter and the various facts depicted in the diagram. The title should be brief, self explannatory, clear and nonambiguous. However, brevity should not be attempted at the cost of clarity, The title should be neatly displayed either at the top of the diagram or at its bottom.

If necessary the footnotes may be given at the left hand bottom of the diagram to explain certain points or facts, not otherwise covered in the title.

3. Selection of Scale

One of the most important factors in the factors in the construction of diagrams is the choice of an appropriate scale. The same set of numercial data if plotted on different scales may give the diagrams differing widely in size and at times might lead to wrong and misleading interpretations. Hence, the scale should be selected with great caution. unfortunately, no hard and fast rules are land down for the choice of scale. As a guiding principle the scale should be selected consistent with the size of the paper and the size of the observations to be displayed so that the diagram and obtained is neither too small not too big. The size of the diagram should be reasonable so as to focus attention on the salient features and important characteristics of the data. The scale showing the values should be in even numbers or multiples of 5 or 10. The scales (s) used on both the horizantal and vertical axes should be clearly indicated. For comparative study of two or more diagrams, the same scale should be adopted to draw valid conclusions.

4. Proportion Between Width and Height

A proper proportion between the dimensions (height and width) of the diagram should be maintained, consistent with the space available. Here again no hard and fast rules are laid down. In this regard Lutz in his book Graphic Presentation has suggested a rule called 'root two' rule, viz, the ratio 1 to $\sqrt{2}$ or 1 to 1.414 between the smaller side and the larger side respectively. The diagram should be generally displayed in the middle (centre) of the page.

5. Choice of a Diagram

A large number of diagrams (discussed below) are used to present statistical data. The choice of a particular diagram to present a given set of numerical data is not an easy one. It primarily depends on the nature of the data, magnitude of the observations and the type of people for whom the diagrams are meant and requires great amount of expertise, skill and intelligence. An inapporpriate choice of the diagram for the given set of data might give a distorted picture of the phenomenon under study and might lead to wrong and fallacious interpretations and conclusions. Hence, the choice of a diagram to present the given data should be made with utmost caution and care.

6. Source Note and Number

As in the case of tables. source note, wherever possible should be appended at the bottom of the diagram. This is necessary as, to the learned audience of Statistics, the reliability of the imformation varies from source to source. Each diagram should also be given a number for ready reference and comparative study.

7. Index

A brief index explaining various types of shades, colours, line and designs used in the construction of the diagram should be given for clear understanding of the diagram.

8. Simplicity

Lastly, digrams should be as simple as possible so that they are easily understood even by a layman who does not have any mathematical or statistical background. If too much imformation is presented in a single complex diagram it will be difficult to grasp and might even become confusing to the mind. Hence, it is advisable to draw more simple diagrams than one or two complex diagrams.

Types of Diagrams :

A large variety of diagrammatic devices are used in practice ot present statistical data. However, we shall discuss here only some of the most commonly used diagrams which may be broadly classified as follows :

(1) One-dimensional diagrams viz, line diagrams and bar diagrams.

(2) Two-dimensional diagrams such as rectangles, square and circles or pie diagrams.

(3) Three-dimensional diagrams such as cubes, sphere, prisms, cylinders and blocks.

(4) Pictograms.

(5) Cartograms.

One-dimensional Diagrams

A. Line Diagram

This is the simplest of all the diagrams. It consists in drawing vertical lines, each vertical line being equal to the frequency. The variate (x) values are presented on a suitable scale along the X-axis and the corresponding frequencies are presented on a suitable scale along Y axis. Line diagrams facilitate comparisons though they are not attractive or appealing to the eye.

Remark : Even a time series data may be presenyted by a line diagram, by taking time factors along X-axis and the variate values along Y-axis.

Problem 1 : The follwoing data shows the number of accidents sustained by 314 drivers of a public utility company over a period of five years.

Number of accidents :

0	1	2	3	4	5	6	7	8	9	10	11

Numbers of drivers :

82	44	68	41	25	20	13	7	5	4	3	2

Represent the data by a line diagram

Solution

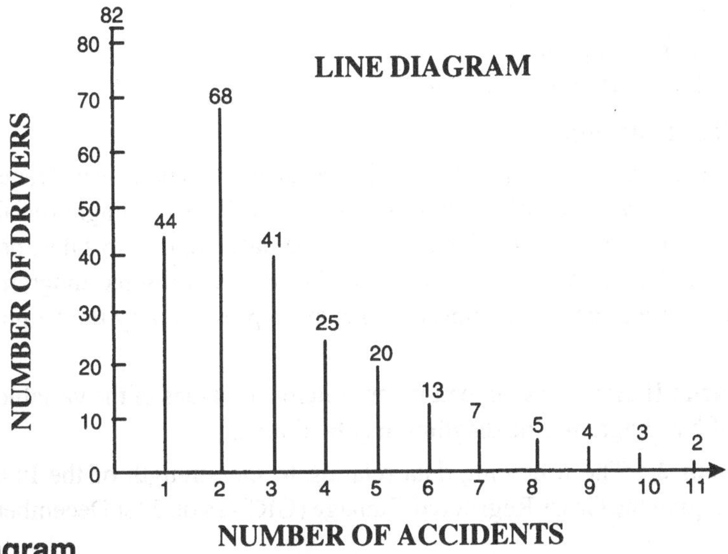

B. Bar Diagram

Bar diagrams are one the easiest and the most commonly used devices of presenting most of the business and economic data. These are especially statisfactory for categorical data or series. They consist of a group of equidistant rectangles, one for each group or category of the data in which the values or the magnitudes are represented by the length or height of the rectangles, the width of the rectangles being arbitrary and immaterial. These diagrams are called one-dimensional because in such diagrams only one demension viz, height (or length) of the rectangles is taken into account to present the given values. The follwoing points may be borne in mind to draw bar diagrams.

(i) All the bars drawn in a single study should be of uniform (though arbitrary) width depending on the number of bars of bars to be drawn and the space available.

(ii) Proper but uniform spacing should be given between different bars to make the diagram look more attractive and elegant.

(iii) The height (length) of the rectangles or bars are taken proportional to magnitude of the observations, the scale being selected keeping in view the magnitude of the largest observation.

(iv) All the bars should be constructed on the same base line.

(v) It is desirable to write the figures (magnitudes) represented by the bars at the top of the bars to enable the reader to have a precise idea of the value without looking at the scale.

(vi) Bars may be drawn vertically or horizontally. However, in practice, vertical bars are generally used because they give and attractive and appealing get up.

(vii) Wherever possible the bars should be arranged from left to right (from top to bottom in case of horizontal bars) in order of magnitude to give a pleasing effect.

Types of Bar Diagrams

The following are the various types of bar diagrams in common use :

(a) Simple bar diagram

(b) Sub-divided or component bar diagram.

(c) Precentage bar diagram.

(d) Multiple bar diagram.

(e) Deviation or Bilateral bar diagram

(a) Simple Bar Diagram

Simple bar diagram is the simplest of the bar diagrams and is used frequently in practice for the comparative study of two or more items or values of a single variable or a single classification or category of data. For exapmpe, the data relating to sales, profits, production, population, etc., for different periods may be presented by bar diagrams. As already pointed out the magnitudes of the observations are represented by the heights of the rectangles.

Remark : If there are a large number of items or values of the variable under study, then instead of bar diagram, line diagram may be drawn.

Problem 2 : The following data relating to the strength of the Indian Merchant Shipping. Fleet give the Gross Registered Tannage (GRT) as on 31st December, for different years.

Year :	1961	1966	1971	1975	1976
GRT in '000:	901	1,792	2,500	4,464	5,115

Source : Ministry of Shipping and Transport

Represent the data by suitable bar diagram.

Solution

STRENGTH OF INDIAN MERCHANT SHIPPING FLEET

GROSS REGISTERED TONNAGE IN '000

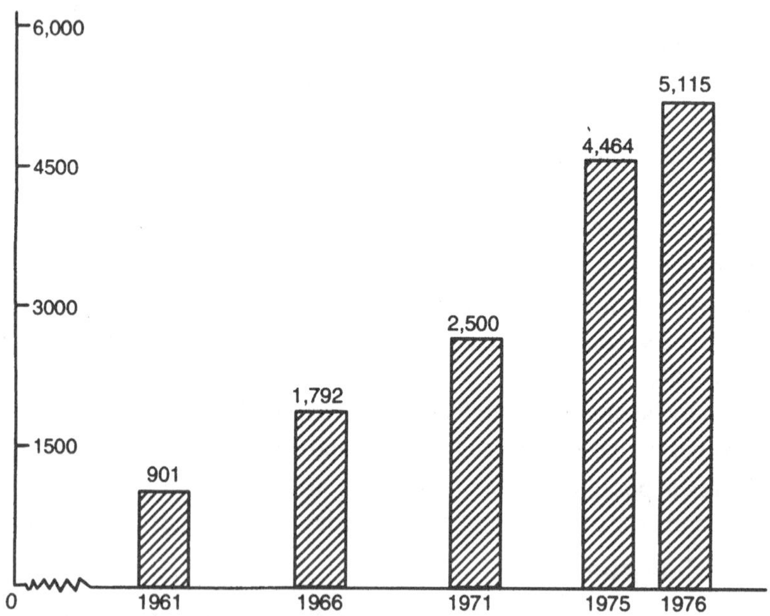

(b) Sub-Divided or Component Bar Diagram

A very serious limitation of the bar diagram is that it studies only one characteristic or classification at a time. For example, the total number of students in a college for the last 5 years can be conveniently expressed by simple bar diagrams but it cannot be used if we have also to depict the faculty-wise or sexwise distribution of students. In such a situation, sub-divided or component bar diagram is used. Sub divided bar-diagrams are useful not only for presenting several items of a variable or a category graphically but also enable us to make comparative study of different parts or components among themselves and also to study the relationship between each component and whole.

In general sub-divided or component bar diagrams are to be used if the total magnitude of the given variable is to be divided into various parts or sub-classes or components. First of all a bar representing the total is drawn. Then it is divided into various segments, each segment representing a given componment of the total. Different shades or colours. crossing or dotting, or designs are used to distinguish the various components and a key or index is given along with the diagram to explain these differences.

In addition to the general rules for constructing bar daigrams, the follwoing points may be kept in mind while constructing sub-divided or component bar diagrams:

(i) To facilitate comparisons the order of the various components different bars should be same. It is customary to show the largest component at the base of the bar and the smallest component at the top so that the various components appear in the order of their magnitude.

(ii) As already pointed, an index or key showing the various components represented by different shades, dotting, colour, etc. should be given.

(iii) The use of sub-divided bar diagram is not suggested if the number of components exceeds 10, because in that case the diagram is loaded with to much information and is not easy to understand and interpret. Pie or circle diagram (discussed later) is appropriate in such a situation. The comparison of the various components in different bar is quite tedious as they do not have a common base and requires great skill and expertise.

Problem 3 : Represent the following data by a suitable diagram :

Items of Expenditure	Family A (Income Rs. 500)	Family B (Income Rs. 300)
Food	150	150
Clothing	125	60
Education	25	50
Miscellaneous	190	70
Saving or Deficit	+10	-30

Solution. The data can be represented by sub-divided bar diagram as shown below :

SUB-DIVIDED BAR DIAGRAM SHOWING EXPENDITURE OF TWO FAMILIES

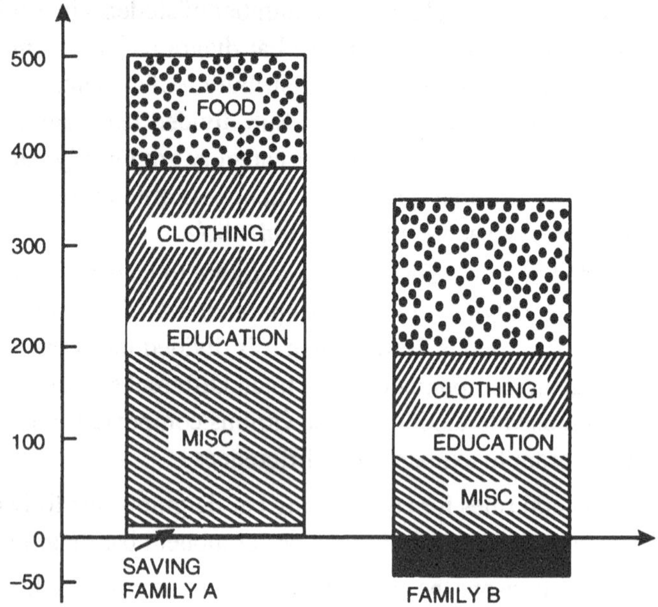

(C) Percentage Bar Diagram

Sub-divided or componment bar diagrams presented graphically on percentage basis give percentage bar diagrams. They are specially useful for the diagrammatic portrayal of the *relative changes* in the data. Percentage bar diagram is used to highlight the relative importance of the various component parts to the whole. The total for each bar is taken as 100 and the value of each component or part is expressed, as percentage of the respective totals. Thus, in a percentage bar diagram, all the bars will be of the same height, viz, 100, while the various segments of the bar representing the different components will varyin height depending on their percentage values to the total. Percentage bars are quite convenient and usefull for comparing two or more sets of data.

Problem 4 : The adjoining table gives the break-upof the expenditure of a family on different itemsof consumption. Draw percentage bar diagramto represent the data.

Item	Expenditure (Rs.)
Food	240
Clothing	66
Rent	125
Fuel and Lighting	57
Education	42
Miscellaneous	190

Solution. First of all we convert the given figures into precentages of the total expenditure as detailed below.

Item	Rs.	Expenditure %	Cumulative %
Food	240	$\frac{240}{720} \times 100 = 33.33$	33.33
Clothing	66	$\frac{66}{720} \times 100 = 9.17$	42.50
Rent	125	$\frac{125}{720} \times 100 = 17.36$	59.86
Fuel and Lighting	57	$\frac{57}{720} \times 100 = 7.92$	67.78
Education	42	$\frac{42}{720} \times 100 = 5.83$	73.61
Miscellaneous	190	$\frac{190}{720} \times 100 = 26.39$	100.00
Total	720	100	

Diagram Showing Expenditure of Family on Different Items of Consumption

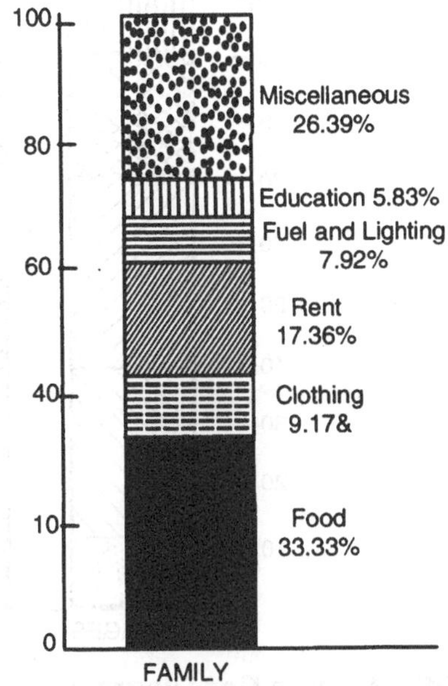

Problem 5 : Draw a bar chart for the following data showing the percentage of total population in villages and towns.

	Percentage of total population in Villages	Towns
Infants and young childern	13.7	12.9
Boys and girls	25.1	23.2
Young men and women	32.3	36.5
Middle-aged men and women	20.4	20.1
Elderly persons	8.5	7.3

Solution :

CALCULATIONS FOR PERCENTAGE BAR DIAGRAMS

Category	Villages		Towns	
	%	Cumulative %	%	Cumulative %
Infants and young children	13.7	13.7	12.9	12.9
Boys and girls	25.1	38.8	23.2	36.1
Young men and women	32.3	71.1	36.5	72.6
Middle aged men and women	20.4	91.5	20.1	92.7
Elderly persons	8.5	100.0	7.3	100.0
	100		100	

PERCENTAGE BAR DIAGRAM SHOWING TOTAL POPULATION IN VILLAGE AND TOWNS BY DIFFERENT CATEGORIES

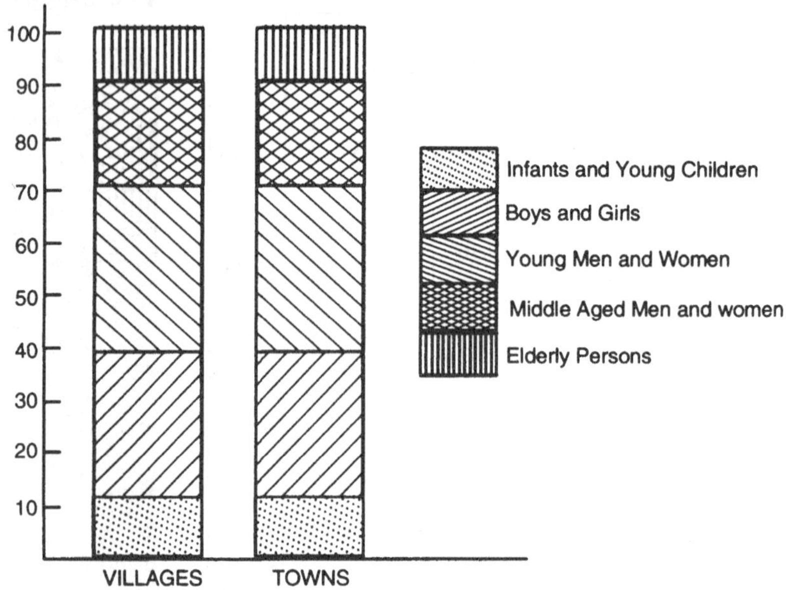

(d) Multiple Bar Diagram

A limitation of the simple bar diagram was that it can be used to portray only a single characcteristic or category of the data. If two or more sets of inter-related pheomena or variables are to be presented graphically, multiple bar diagrams are used. The technique of drawing multiple bar diagram is basically same as that of drawing simple bar diagram. In this case, a set of adjacent bars (one for each variable) is drawn. Proper and equal spacing is given between different sets of the bars. To distinguish between the different bars in a set, different colours, shades, dottings or crossings may be used and key or index to this effect may be given.

Problem 6 : The data below give the yearky profits (in thousand of rupees) of two companies A and B

Year	Profits in ('000 rupees)	
	Company A	**Company B**
1994-95	120	90
1995-96	135	95
1996-97	140	108
1997-98	160	120
1998-99	175	130

Represet the data by means of a suitable diagram.

Solution : The data can be suitably represented by a multiple bar diagram as shown below.

YEARLY PROFITS IN '000 RUPEES

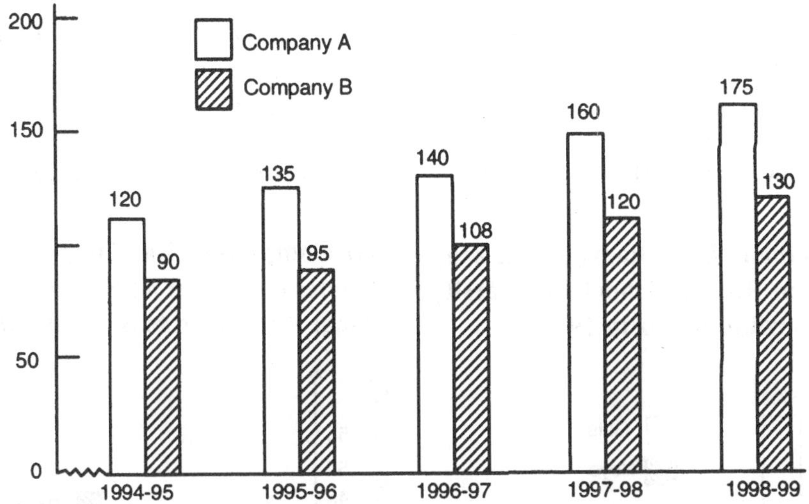

Remark : A careful examination of the above figures of profits for the two companies A and B reveals that in all the years from 1994-95 to 1998-99, company A shows higher profits than the company B. In such a situation when the values of one concern or unit show an increase over the values of the other concern or unit for all the periods under consideration, the data can be elegantly represented by a special type of subdivided bar diagram, in which total refers to the values of the concern or unit with higher values and the lower portion (shaded) of the bar shows the values of other concern. The remaining portion (blank) shows the balance (excess) of the two concerns or units. We represent below the above data in this manner.

YEARLY PROFITS IN '000 RUPEES

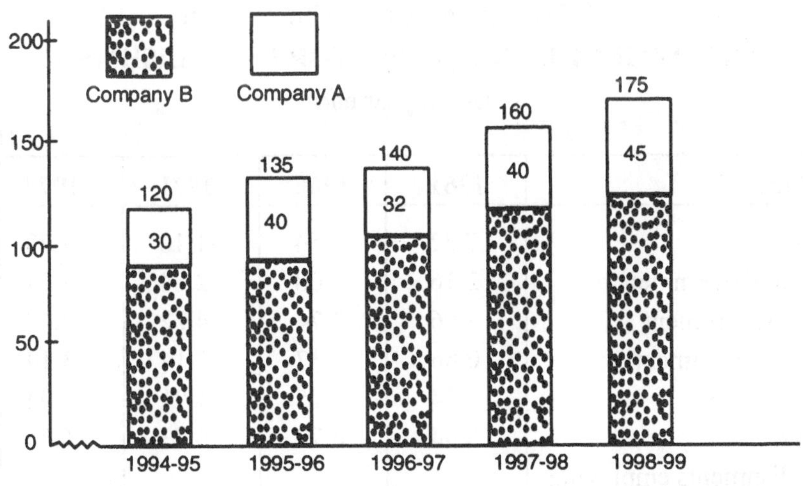

Problem 7 : The following data shows she students in millions on rolls at school/university stage in India according fo different class groups and sex for the year 1970-71 as an 31st March.

Stage	Boys	Girls	Total
Class 1 to V	35-74	21.31	57.05
Class VI to VIII	9.43	3.89	13.32
Class IX to XI	4.87	1.71	6.58
University/College	2.17	0.64	2.81

Represent the data by (i) Component bar diagram and (ii) Multiple bar diagram.

Solution

(i) Component Bar Diagram Showing Students on Roll At School / University Stage According to Sex in 1970-71

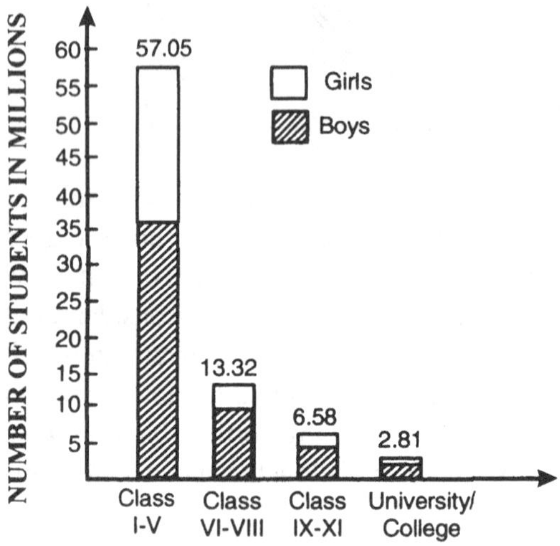

(i) Component Bar Diagram Showing Students on Roll At School / University Stage According to Sex in 1970-71

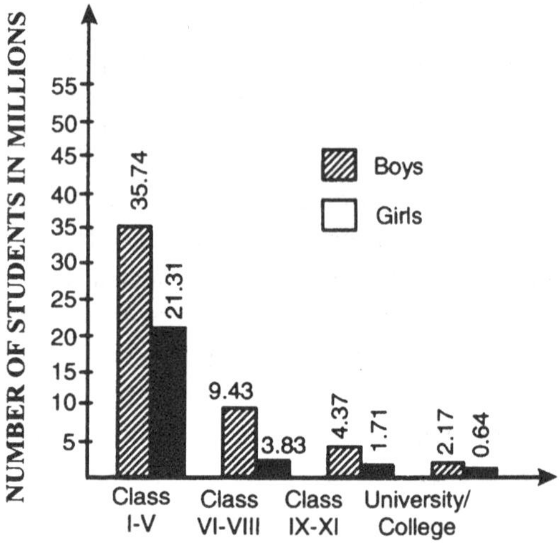

Problem 8 : Represent the following data relating employment in public and private sector establishments in India for different years by a suitable diagram.

EMPLOYMENT IN PUBLIC SECTOR ESTABLISHMENTS

(As at year end)

(In million)

Sector	1961	1966	1971	1974	1975
PUBLIC	7.32	9.54	11.10	12.73	13.13
Central Government	2.16	2.68	2.84	2.97	3.00
State Government	3.06	3.74	4.22	4.68	4.83
Quasi Government	0.86	1.37	2.14	3.14	3.32
Local Bodies	1.24	1.75	1.91	1.94	1.98
PRIVATE	5.04	6.72	6.73	6.75	6.79
Establishments employing					
25 or more persons	5.04	6.01	6.02	6.03	6.08
10 to 24 persons	-	0.71	0.72	0.72	0.71
Total	12.36	16.26	17.83	19.48	19.92

Source : Directions General of Employment and Training.

Solution : Here, we represent the data by Sub-divided multiple bars as shown below :

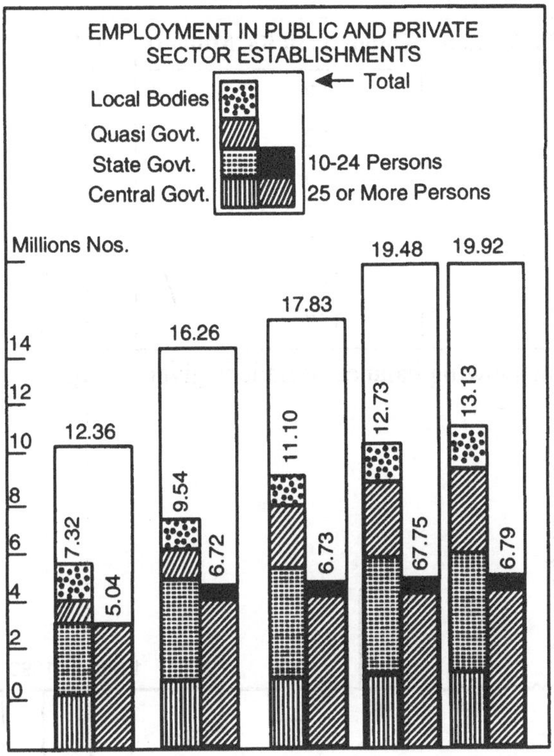

(e) Deviation Bars

Deviation bars are specially useful for graphic presentation of net quantities viz, surplus or deficit, e.g. net profit or loss, net of imports and exports which have both positive and negative values. The positive deviation (e.g., profits. surplus) are presented by bars above the base line while negative deviations (loss deficit) are represented by bars below the base line. The following example will illustrate the points.

Remark : Deviation bars are also sometimes known as Bilateral Bar Diagrams and are used to Depict plus (surplus) and minus (deficit) directions from the point of reference.

Problem 9 : For the following data prepare a suitable diagram showing Balance of Trade:

Year	Exports (In Rs. Million)	Inports (In Rs. Million)
1994	24	9
1995	115	92
1996	84	92
1997	110	120
1998	130	183
1999	162	187

Solution :

Year	Exports (In Rs. Million)	Imports (In Rs. Million)	Balance of Trade (In Million Rs.)
1994	24	9	24 - 9 = 15
1995	115	92	115 - 92 = -8
1996	84	92	84 - 92
1997	110	120	110 - 120 = -10
1998	130	183	130 - 183 = -53
1999	162	187	162 - 187 = -25

Deviation bar diagram showing balance of trade is given in Fig.

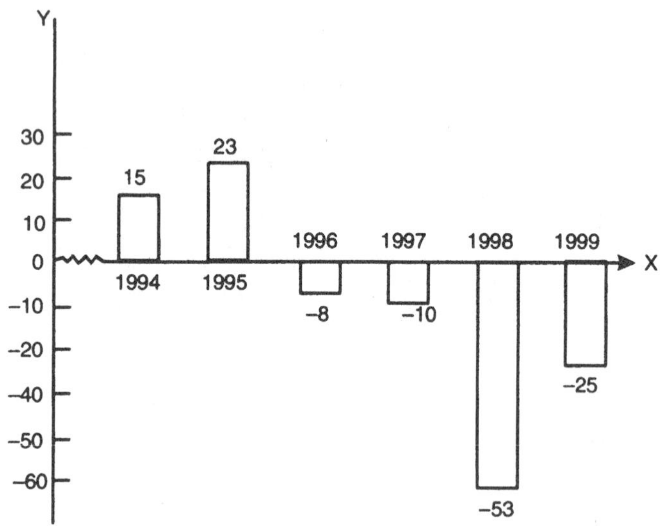

(F) Broken Bars

Broken bars are used for graphic presentation of the data which contain very wide variations in the values i.e., the data which contain very large observations along with small observations. In this case the squeezing of the vertical scale will not be of much help because it will make the small bars to look too small and clumsy and thus will not reveal the trune characteristics of the data. In order to provide adequate and reasonable shape for smaller bars, the larger (or largest) bar (s) may be broken at the top.

Remark : However, if all the observations are fairly large so that all the bars have a broken vertical axis, then instead they can be drawn with a false base line for the vertical axis.

Problem 10 : The following data relates to the imports of foreign merchandise and exports (including re-exports) of Indian merchandise in million rupees) for some countries for the year 1975-76.

Country	Imports	Exports
Burma	53	89
Czecholovakia	522	343
Canada	2,278	424
Australia	1,015	477
Italy	799	785
France	1,852	835
Germany (F. R.)	3.566	1,173
Iran	4.593	2,708
United Kingdom	2,683	4,020
USSR	2,958	4,128
Japan	3,548	4,263
USA	12,699	5,054

Represent the data by suitable diagram.

Solution : The above data is represented by bar-diagrams as shown below.

FOREIGN TRADE BY COUNTRIES
1975-76

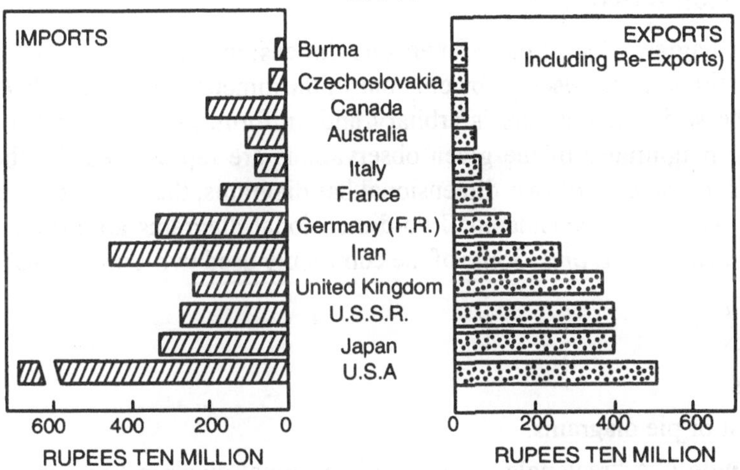

Problem 11 : Represent the following data relating to the military statistics at the border during the war between the two counrties A and B in 1999 by multiple bar diagram.

Category	Country A	Country B
Army Divisions	4	20
Semi-Army Units	50	-
Fighter Planes	75	700
Tanks	50	300
Total Troops	100,000	170,000

Solution :

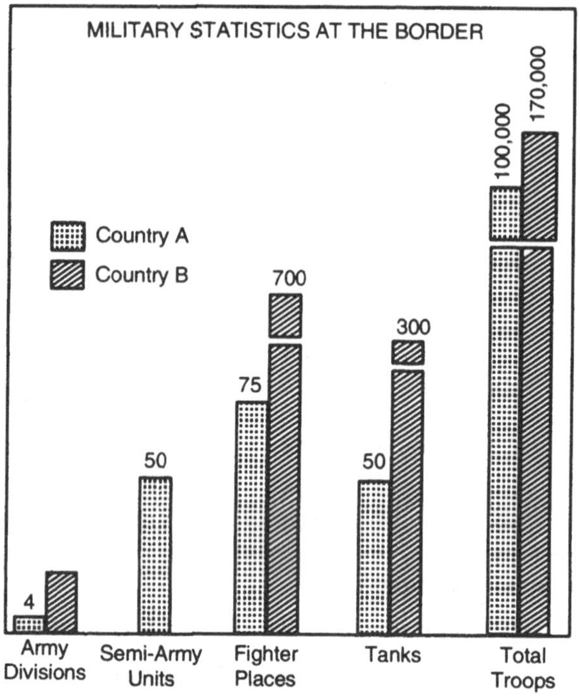

Two-dimensional Diagrams

Line or bar diagrams discussed so far are one-dimensional diagrams since the magnitudes of the observations are represented by only one of the dimensions viz., height (length) of the bars while the width of the bars is arbitary and uniform. However, in two-dimensional diagrams, the magnitudes of the given observations are represented by the area of the diagram, This, in the case of two-dimensional bar diagrams, the length as well as width of the bars will have to be considered. Two-dimensional diagrams are aslo known as *area diagrams or surface diagrams*. Some of the commonly used two-dimensional diagrams are

(A) Rectangle

(B) Squares

(C) Circles

(D) Angular or pie diagrams.

(A) Rectangle : A "rectangle" is a two-dimensional diagram because it is based it is based on the area principle. Since the area of a rectangle is given by the product of its length and breadth, in a rectangle diagram both the dimensions viz., length (height) and width of the bars is taken into consideration.

Just like bars, the rectangles are placed side by side, proper and equal spacing being given between different rectangles. In fact, rectangle diagrams are a modified form of bar diagrams and give a more detailed information than bar diagrams.

Like sub-divided bars, we have also sub-divided rectangles for depicting the total and its break-up into various components. Likewise percentage rectangle diagram may be used to portray the relative magnitudes of two or more sets of data and their components making up the total. We give below a few illustrations.

Problem 12 : Prepare a rectangular diagram from the following particulars relating to the production of a commodity in a factory.

Units produced	1,000
Cost of raw materials	Rs. 5,000
Direct expenses	Rs. 1,000
Indirect expenses	Rs. 1,000
Profit	Rs. 1,000

Solution : First of all we will find the cost of material , expenses and profits per unit as given below :

$$\text{Cost of raw material per unit} = \text{Rs.}\frac{5000}{1000} = \text{Rs. } 5$$

$$\text{Direct expenses per unit} = \text{Rs.}\frac{2000}{1000} = \text{Rs. } 2$$

$$\text{Indirect expenses per unit} = \text{Rs.}\frac{1000}{1000} = \text{Re. } 1$$

$$\text{Profit per unit} = \text{Rs.}\frac{1000}{1000} = \text{Re. } 1$$

Diagram Showing Cost and Profit For A Commodity in a Factory

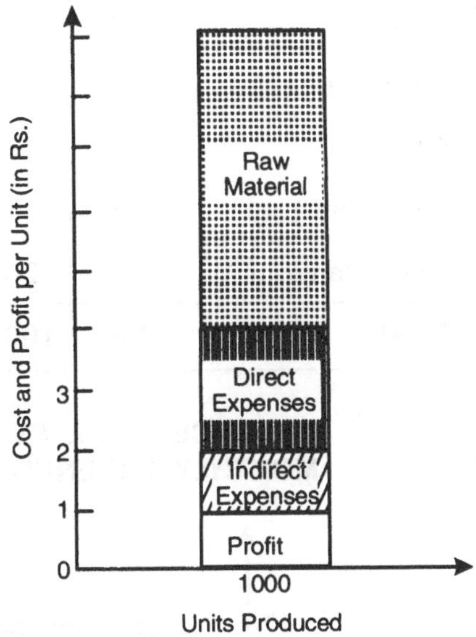

Problem 13 : The following data relates to the monthly expenditure (in Rs.) of two families A and B.

Item of Expenditure	Expenditure (in Rs.)	
	Family A	Family B
Food	160	120
Clothing	80	32
Rent	60	48
Light and fuel	20	16
Miscellaneous	80	24
Total	400	240

Represent it by a suitable percentage diagram.

Solution :

Since the total expenses of the two families are different, an appropriate percentage diagram for the above data will be rectangular diagram on percentage basis. The percentage bar diagram will not be able to reflect the inherent difference in the total expenditures of the two families.

The widths of the rectangeles will be taken in the ratio of the total expenses of the two families viz., 400 : 240 i.e., 5 : 3.

CALCULATIONS FOR PERCENTAGE RECTANGULAR DIAGRAM

Item of Expenditure	Family A			Family B		
	Rs	%	Cumulative%	Rs.	%	Cumulative %
Food	160	40	40	120	50	50
Clothing	80	20	60	32	13.33	63.33
Rent	60	15	75	48	20	83.33
Light and fuel	20	5	80	16	6.67	90
Miscellaneous	80	20	100	24	10	100
Total	400	100		240	100	

PERCENTAGE RECTANGLE DIAGRAM SHOWING MONTHLY EXPENDITURE OF TWO FAMILIES A AND B

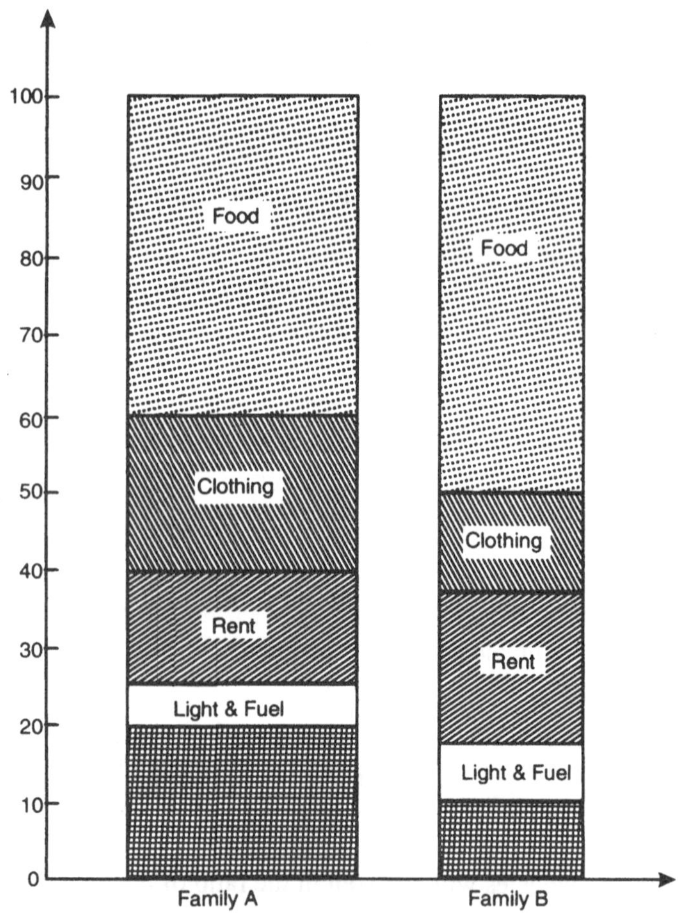

Problem 14 : Represent the following data by a percentage sub divided bar diagram

Item of Expenditure	Family A	Family B
Food	150	150
Clothing	125	60
Education	25	50
Miscellaneous	190	70
Savings or Deficit	+10	-30

Solution : Since the total incomes of the two families are different, an appropriate percentage bar diagram for the above data will be rectangular diagram on percentage basis. The percentage bar diagram will not be able to reflect the inherent differences in the total incomes in the two families.

The widths of the rectangles will be taken in the ratio of the total incomes of the families viz., 500 : 300 i.e., 5 : 3.

CALCULATIONS FOR PERCENTAGE RECTANGULAR DIAGRAM

Item of Expenditure	Family A			Family B		
	Expenditure Rs	%	Cumulative %	Expenditure Rs	%	Cumulative %
Food	150	30	30	150	50	50
Clothing	125	25	55	60	20	70
Education	25	5	60	50	16.7	86.7
Miscellaneous	190	38	98	70	23.3	110.0
Savings or Deficit	+10	2	10	–30	–10	100
Total	500			300		

PERCENTAGE DIAGRAM SHOWING MONTHLY INCOME AND EXPENDITURE OF TWO FAMILIES A AND B

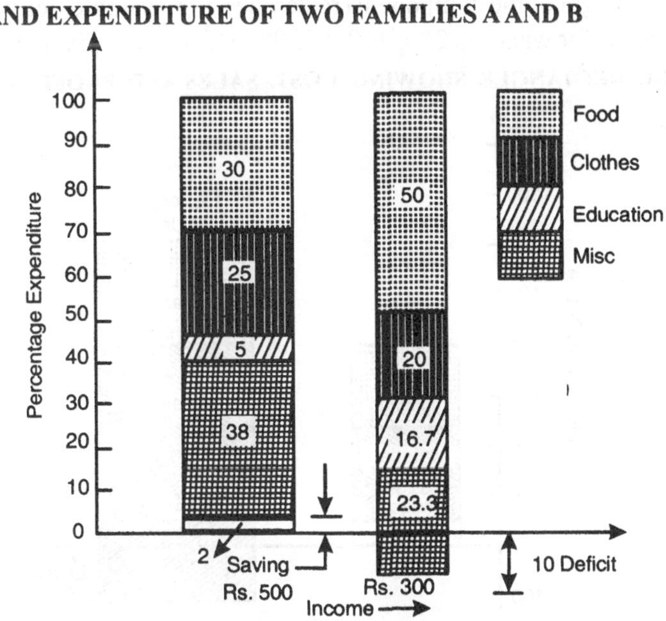

Problem15 : Draw a suitable diagram to represent the following information.

	Selling price Per unit (in Rs.)	Quantity sold	Total Cost (in Rs.)			
			Wages	Materials	Misc	Total
Factory X	400	20	3,200	2,400	1,600	7,200
Factory Y	600	30	6,000	6,000	9,000	21,000

Show also the profit or loss as the case may be

Solution : First of all we shall calculate the cost (wages, materials, misc,) and profit per unit as given in the following table.

	Selling price per unit (in Rs.)	Quantity sold	Cost per unit (in Rs.)				Profit per unit (in Rs.)
			Wages	Material	Misc.	Total	
Factory X	400	20	160	120	80	360	400-360=40
Factory Y	600	30	200	200	300	700	600-700=100

Note : Negative profit is regarded as loss.

An appropriate diagram for representing this data would be the 'Rectangles' whose width are in the ratio of the quantities sold i.e., 20 : 30 i.e., 2:3. Selling price would be represented by the corresponding heights of the rectangles with various factors of cost (wages, material, misc) and profit or loss represented by the various divisions of the rectangles as shown in the following diagram.

Remark : In the case of profit i.e., when selling price (S.P.) is greater than cost price (C.P.), the entire rectangle will lie above the X-axis, the segment just above the X-axis showing profit. But in case of loss i.e., when S.P. is less than C.P., we will have the rectangle with a portion lying below the X-axis which will reflect the loss incurred i.e., the cost not recovered through sales. The values of each component are given by the product of the base with the corresponding height of the component (rectangle). For example, for the factory X, the area of the component for wages is 20 x 160 = 3200, which is the given cost.

SUB-DIVIDED RECTANGLE SHOWING COST, SALES AND PROFIT OR LOSS PER UNIT

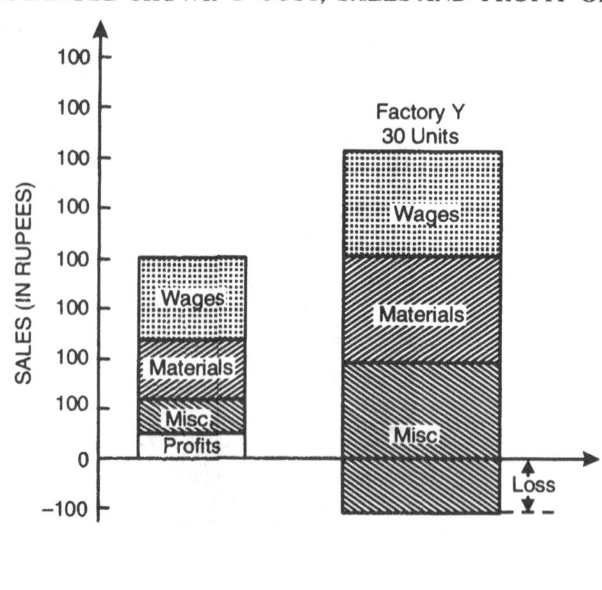

Aliter. The above data can also be expressed as follows :

	Factory X	Factory Y
Total sales (in Rs.) Total cost (in Rs.)	400 x 20 = 8,000 7,200	600 x 30 = 18,000 21,000
Profit (in Rs.)	800	- 3,000

Note : Negative profit is regarded as loss.

Data are now expressed diagrammatically by sub-divided (or component) bar diagram, by taking the total sales as bars along the Y-axis and various factors of cost (viz., wages, materials, miscellaneous) and profit as the various divisions of the bars as shown on in Fig.

SUB-DIVIDED BAR DIAGRAM SHOWING COST. SALES AND PROFIT OR LOSS

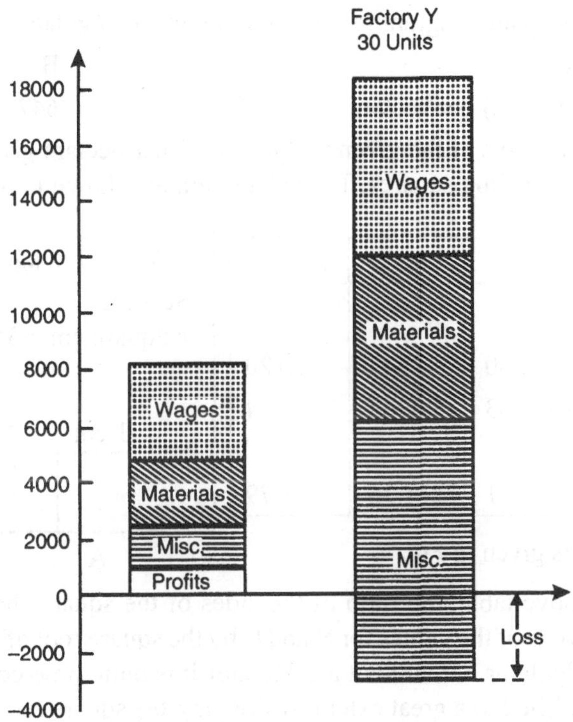

(B) Square Diagrams

Among the two-dimensional diagrams, squares are specially useful if it is desired to compare graphically the values or quantities which differ widely from one another as e.g., the population of different countries at a given time or of the same country at different times or the imports or exports of different countries. In such a situation, the bar diagrams are not suitable, since they will give very disproportionate bars i.e., the bars corresponding to smaller quantities would be comparatively too small and those corresponding to bigger values would be too big. In particular, if two values are in the proportion of 1 : 25 and if we draw bar diagram, on (bigger) will be 25 times (in height) than that of the other (smaller). In such a situation, square diagrams give a better presentation.

Like rectangle diagram, square diagram is a two-dimensional diagram in which the given values are represented by the area of the square. Since the area of the square is given by the square of its side, the side of the square diagram will be in proportion to the square root of the given observations. Thus if the two observations are in the ratio of 1:25, the sides of the squares will be in the ratio of their square roots viz., 1:5.

Construction of the square diagrams is quite simple. First of all we obtain the square roots of the given observations and then squares are drawn with sides proportional to these square roots, on an appropriate scale which must be specified.

Remark 1 : The square may be drawn horizontally (on the same base line) or vertically one below the other to facilitate comparisons. However, in practice, the first method viz, horizontal presentation is generally used since if economises space.

2 : Although square diagram is a two-dimensional diagram, it is used to depict only a single magnitude or value.

Problem16 : Draw a square diagram to represent the following data.

Country	A	B
Yield in (kg.) per hectare	350	647

Solution : The square roots of the given yields in (kg) per hectare give the porportuion of the sides of the corresponding squares. The calculation are shown in the following table :

Country	A	B	C
Yield in kg per hectare	350	647	1,120
Square root	18.7083	25.4362	33.4664
Ratio of the sides of the square	1	1.36	1.79

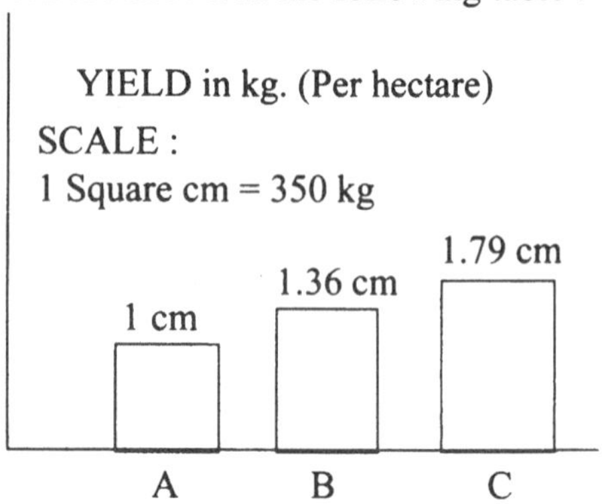

The square diagram is given in Fig.

Remark : In the above table, the ratio of the sides of the squares has been obtained on dividing the square roots of the values for B and C by the square root of the value for A. This is easy to do on a calculator but without a calculator it is quite time-consuming. However, the thing can be simplified to a great extent by dividing the square roots of the values of A, B and C by a whole number, say, 15 or 18. Division by, say, 15, gives the ratio of the sides of square for A, B and C as 1.25, 1.70, 2.23 respectively. Squares can now be constructed by taking appropriate scale which must be specified in the diagram.

If we construct squares with sides 1.25 cms., 1.70 cms., and 2.23 cms., the scale will be obtained as follows :

Area of square for A is $(1.25)^2 = 1.5625$ square cms. This area represents the value 350 kg. Thus, follows:

$$1.5625 \text{ sq. cms.} = 250 \text{ kg.} \quad \Rightarrow \quad 1. \text{ sq. cm.} = \frac{350}{1.5625} = 224 \text{ kg}$$

(C) Circle Diagram

Circle diagrams are alternative to suqare diagrams and are used for the same purpose, viz.., for diagrammatic presentation of the values differing widely in their magnitude. The area of the circle, which represents the given values is given by πr^2, where $\pi = 22/7$ and r is the radius of circle. In other words, the area of the circle is proportional to the square of its radius and consequently, in the construction of the circle diagram the radius of the circle is a value proportional to the square root of a given magnitude. Accordingly, the lengths which were taken as the sides of the square may also be taken as the radil of the circles representing the given magnitudes.

Remark

Circle diagrams are more attractive and appealing than square diagrams and since both require more or less the same amount of work, viz., computing the square roots of the given magnitudes(rather circles are easy to draw). circle diagrams are generally preferred to square diagrams.

Since square and circle diagrams are to be compared on an area basis, it is difficult to judge the background. Accordingly, proper care should be taken in interpret them. They are also more difficult to construct than the rectangle diagrams.

Scale : The scale to be used for constructing circle diagrams can be calculated as follows : For a given magnitude' 'a' we have

$$\text{Area } \pi r^2 \text{square units} = a \Rightarrow 1 \text{ square} = \frac{a}{\pi r^2}$$

Problem 17. Represent the data of Problem 16 by a circular diagram.

Solution : The data of Example 16 can be represented by a circular diagram on taking the lengths of the sides of the squares which were taken in Example 16, as the radii of the corresponding circle. However, in this case, the scale will be modified accordingly.

$$\text{Scale} : 1 \text{ sq.cm.} \frac{350}{\pi} = \frac{2450}{22} = 111.36 \text{ kg}$$

YIELD IN KG (PER HECTARE)

Scale : 1 sq. cm. = 111.36 kg

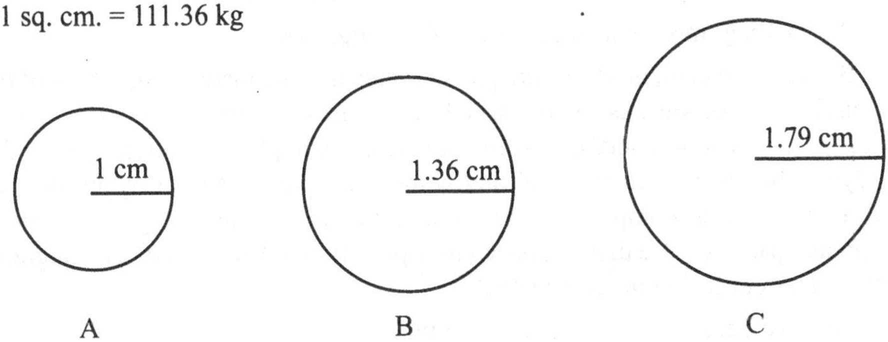

(D) Angular or Pie Diagram

Just as sub-divided and percentage bars or rectangles are used to represent the total magnitude and its various components, the circle (representing the total) may be divided component parts to the total. Such a sub-divided circle diagram is known as an **angular or pie diagram**, named so because the various segments resemble slices cut from a pie.

Step for Construction of Pie Diagram

1. Express each of the component values as a percentage of the repective total.
2. Since the angle at the centre of is 360^0, the total magnitude of the various components is taken to be equal to 360^0 and each component part is to be expressed proportionately in degrees. Since 1 percent of the total value is equal to $360/100 = 3.6^0$, the percentage of the component parts obtained is step I can be converted to degrees by multiplying each of them by 3.6.
3. Draw a circle of appropriate radius using an appropriate scale depending on the space avilable. If only one category or charracteristic is to be used, the circle may be drawn of any radius. However, if two or more sets of data are to be presented simulataneously for comparative studies, then the radii of the corresponding circles are to be proportional to the square roots of their total magnifudes.
4. Having drawn the circle, draw any radius (preferably horizontal). Now with this radius as the base line draw an angle at the centre [with the help of protractor (D)] equal to the degree represented by the first component, the new line drawn at the centre to from this angle will touch the circumference. The sector so obtained will represent the proportion of the first component. From this second line as base, now draw another angle at the centre equal to the degree represented by the 2nd component, to give the sector representing the proportion of the second component. Proceeding similarly, all the sectors representing different component parts can be constructed.
5. Different sectors representing various component parts should be distinguished from one another by using different shades, dottings colour, etc., or giving them explanatory or descriptive labels either inside the sector (if possible or just outside the circle with proper indentification.

Remarks 1: The degrees represented by the various component parts of a given magnitude can be obtained directly without computing their precentage to the total value as follows :

$$\text{Degree of component part} = \frac{\text{Component Value}}{\text{Total value}} \times 360°$$

2. Pie diagrams are also called circular diargrams.

3. Since the comparison of the pie diagrams is to be made on the basis of the areas of the circles and of various sectors which are difficult to be ascertained visually with precision, generally sub-divided or percentage bars or rectangles are preferred to pie diagrams for studying the changes in the total and component parts. Moreover, pie diagrams are difficult to construct as compared with bars or rectangle diagrams. Any way, if the number of component parts is more than 10, pie chart is preferred to bar or rectangle diagram which becomes rather confusing in such a situation.

We give below some illustrations of pie diagrams.

Problem 18 : Draw a pie diagram to represent the following data of proposed expenditure by a State Government for the year 1997-98

Items	Agriculture & Rural Development	Industries & Urban Development	Health & Education	Miscellaneous
Proposed Expenditure (in million Rs.)	4,200	1,500	1,000	500

Solution :

CALCULATIONS FOR PIE CHART

Items (1)	Proposed expenditure (in million Rs.) (2)	Angle at the centre $(3) = \frac{(2)}{7200} \times 360^\circ$
Agriculture and Rural Development	4,200	$\frac{42}{72} \times 360^\circ = 210^\circ$
Industries and Urban Development	1,500	$\frac{15}{72} \times 360^\circ = 75^\circ$
Health and Education	1,000	$\frac{10}{72} \times 360^\circ = 50^\circ$
Miscellaneous	500	$\frac{5}{72} \times 360^\circ = 25^\circ$
Total	7,200	360°

PIE DIAGRAM REPRESENTING PROPOSED EXPENDITURE BY STATE GOVERNMENT ON DIFFERENT ITEMS FOR 1997-98

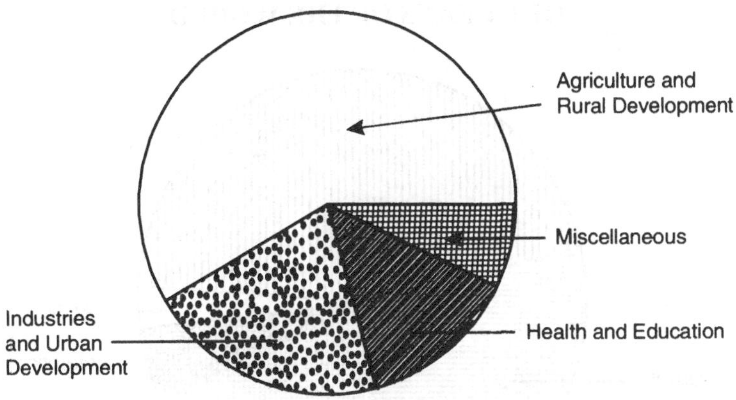

Problem 19 : The following table shows the area in millions of sq. km. of oceans of the world:

Ocean	Area (million sq. km.)
Pacific	70.8
Atlantic	41.2
Indian	28.5
Antarctic	7.6
Arctic	4.8

Draw a pie diagram to represent the data.

Solution

CALCULATIONS FOR PIE DIAGRAM

Occean	Area (million sq. km.)	Angle at the centre
Pacific	**70.8**	$\frac{70.8}{152.9} \times 360^\circ = 166.7^\circ$
Atlantic	41.2	$\frac{41.2}{152.9} \times 360^\circ = 97.0^\circ$
Indian	28.5	$\frac{28.5}{152.9} \times 360^\circ = 67.1^\circ$
Antarctic	7.6	$\frac{7.6}{152.9} \times 360^\circ = 17.9^\circ$
Arctic	4.8	$\frac{4.8}{152.9} \times 360^\circ = 11.3^\circ$
Total	152.9	360°

PIE DIAGRAM SHOWING THE AREA (IN MILLIONS OF SQUARE KMS) OF OCEANS OF THE WORLD

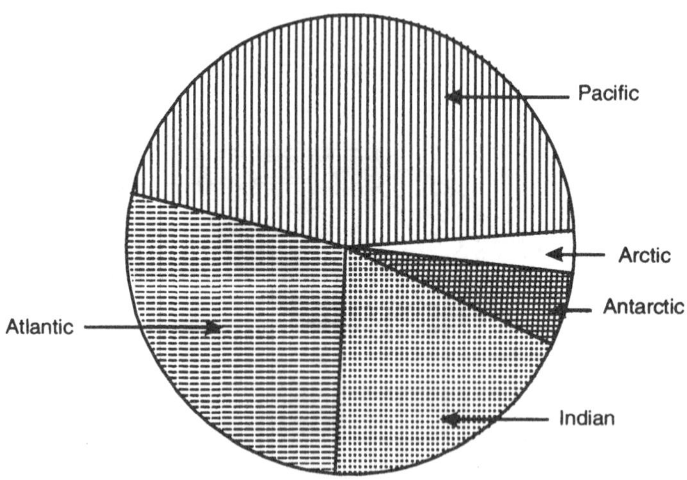

Problem 20 : The following data shows the expenditure on various heads in the first three five year plans (in crores of rupees).

Subject	Expenditure (in crores Rs.)		
	First Plan	Second Plan	Third Plan
Agriculture and C.D.	361	529	1068
Irrigation and Power	561	865	1662
Village and Small Industries	173	176	264
Industry and Minerals	292	900	1520
Transport and Communications	497	1300	1486
Social Services and Miscellaneous	477	830	1500
Total	2361	4600	7500

Represent the data by angular (pie) diagrams.

Solution :

CALCULATIONS FOR PIE DIAGRAMS

Items of Expenditure	Expenditure (in crores Rupees)					
	First Plan		Second Plan		Third Plan	
	Rs.	Degrees	Rs.	Degrees	Rs.	Degrees
Agriculture and C.D.	361	$\frac{361}{2361} \times 360° = 55.1°$	529	$\frac{529}{4600} \times 360° = 41.4°$	1.068	$\frac{1068}{7500} \times 360° = 51.2°$
Irrigation and power	561	$\frac{561}{2361} \times 360° = 85.5°$	865	$\frac{865}{4600} \times 360° = 67.7°$	1,662	$\frac{1662}{7500} \times 360° = 79.8°$
Village and Small Industries	173	$\frac{173}{2361} \times 360° = 26.4°$	176	$\frac{176}{4600} \times 360° = 13.8°$	264	$\frac{264}{7500} \times 360° = 12.7°$
Industry and Minerals	292	$\frac{292}{2361} \times 360° = 44.5°$	900	$\frac{176}{4600} \times 360° = 13.8°$	1,520	$\frac{1520}{7500} \times 360° = 73.0°$
Transport and Comm-unications	497	$\frac{497}{2361} \times 360° = 75.8°$	1300	$\frac{1300}{4600} \times 360° = 101.7°$	1,486	$\frac{1486}{7500} \times 360° = 71.3°$

Social Services and Miscellaneous	477	$\frac{477}{2361} \times 360° = 72.7°$	830	$\frac{830}{4600} \times 360° = 65.0°$	1,500	$\frac{1500}{7500} \times 360° = 72.0°$
Total	2,361	360°	4,600	360°	7,500	360°
Square Root	48.59		6782		86.60	
Radii of circles	1.0		1.4		1.8	

EXPENDITURE ON VARIOUS HEADS IN FIRST THREE FIVE-YEAR PLANS

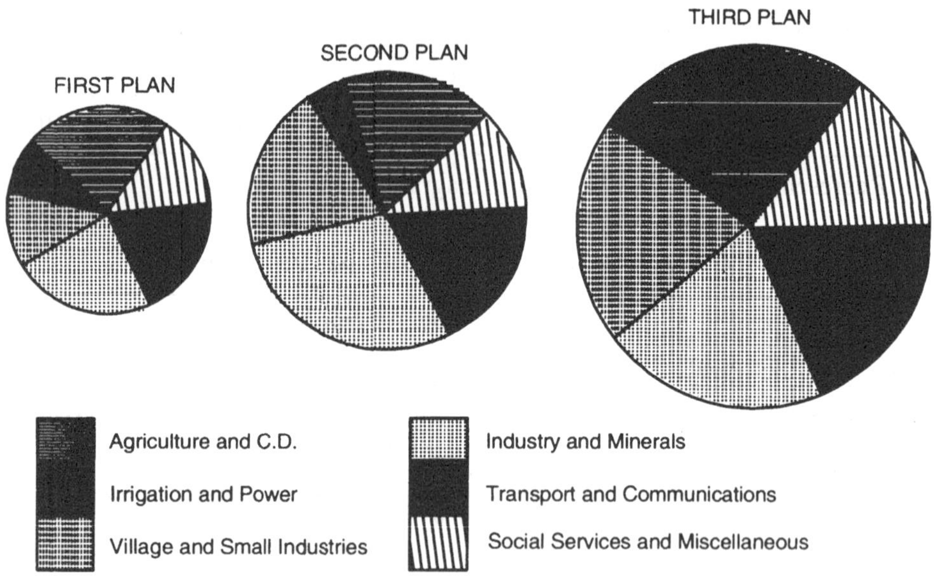

Three - Dimensional Diagrams :

Three-dimensional diagrams, also termed as *volume* diagrams are those in which three dimensions, viz., length, breadth and height are taken into account. They are constructed so that the given magnitudes are represented by the volumes of the corresponding diagrams. The common forms of such diagrams are cubes, spheres, cylinders, blocks, etc. These diagrams are specially useful if there are very wide variations between the smallest and the largest magnitudes to be represented. Of the various three-dimensional diagrams, 'cubes' are the simplest and most commonly used devices of diagrammatic presentation of the data.

Cubes : For instance, if the smallest and the largest magnitudes to be presented are in the ratio of 1 : 1000, the bar diagrams cannot be used because the height of the biggest bar would be 1000 times the height of the smallest bar and thus they would look very disproportionate and clumsy. On the other hand, if square or circle diagrams are used then the sides (radii) of the squares (circles) will be in the ratio of the square roots viz., 1 : $\sqrt{1000}$ i.e., 1:31-63 i.e. 1:32 (approx.), which will again give quite disproportionate diagrams. However, if cubes are used to present this data, then since the volume of cube of side χ is χ^3, the sides of the cubes will be in the ratio of their cube roots viz., 1 $\sqrt[3]{1000}$ i.e. 1 : 10, which will give reasonably proportionate diagrams as compared to one-dimensional or two-dimensional diagrams.

Construction of a Cube a Side 'χ'. The various steps are outlined below :

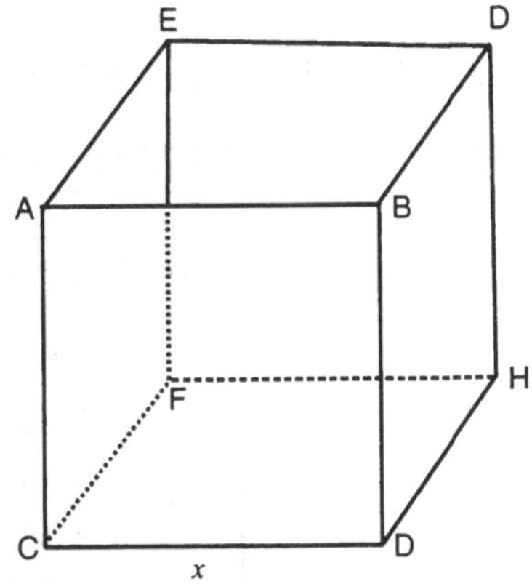

(i) Construct a square ABCD of side χ.

(ii) Draw EF as right bisector i.e., perpendicular bisector of AB, (This is done by finding the mid-point of AB and then drawing perpendicular at that point to the line AB) such that EF = AB and half of it is above AB and half of it is below AB.

(iii) Join AE, CF and EF.

(iv) Through B draw a line BG parallel to AE (i.e., BG II AE) such that BG = AE.

(v) Join EG and through G draw a line GH II EF such that GH = EF.

(vi) Join D and H.

(vii) Rub off the lines CF, EF and FH. Now CDHGEAC is the required cube.

Remarks 1: As already discussed, three-dimensional diagrams are used with advantage over one or two-dimensional diagrams if the range i.e., the gap between the smallest and the largest observation to be presented, is very large. Moreover, they are more beautiful and appealing to the eye than bars, rectangles, squares or circles. However, since three-dimensional diagrams are quite difficult to construct and comprehend as compared to one-or two-dimensional diagrams, they are not very popular. Further, as the magnitudes are represented by the volumes of the cubes (volumes of the three-dimensional diagrams, in general), it is very difficult to visualise and hence interpret them with precision.

2. Cylinders, spheres and blocks are quite difficult to construct and are, therefore, not discussed here.

3. It is worthwhile pointing out here that now-a-days projection techniques are used to represent even one-dimensional diagrams as three-dimensional diagrams for giving them a beautiful and attractive get up.

The following table gives the population of India on the basis of religion.

Religion	Hinduism	Islam	Christianity	Sikhism	Others
Number (in lakhs)	2031.9	354.0	81.6	62.2	36.3

Represent the data by cubes.

Solution : The sides of the cubes will be proportional to the cube roots of the magnitudes they represent.

Note : To compute cube root of a, viz, $\sqrt[3]{a}$ or $a^{1/3}$, let $y = \sqrt[3]{a} = a^{1/3}$

Taking logarithm of both sides : $\log y = \frac{1}{3} \log_{10} a \Rightarrow y = \text{Antilog}\left[\frac{1}{3} \log_{10} a\right]$

COMPUTATION OF CUBE ROOTS

A	$\log_{10}$ a	$\frac{1}{3}\log_{10}$a	Antilog $\left[\frac{1}{3}\log_{10} a\right]$	Ratio of sides
2031.9	3.3079	1.1026	12.67	3.83 ≃ 3.8
354.0	2.5441	0.8480	7.047	2.13 ≃ 2.1
81.6	1.9117	0.6372	4.337	1.31 ≃ 1.3
62.2	1.7938	0.5979	3.962	1.19 ≃ 1.2
36.3	1.5599	0.5199	3.311	1

Now we can express the data diagrammatically by drawing cubes with sides proportional to the values given in the last column.

Scale : $\frac{1}{8}$ cms³ = 36.3 lakhs

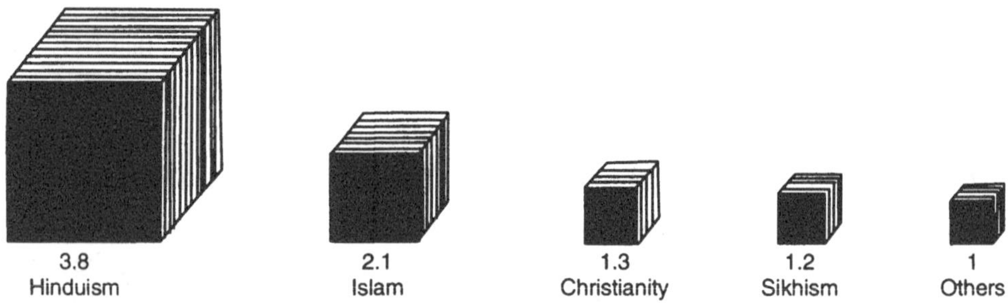

Pictograms :

Pictograms is the technique of presenting statistical data through appropriate pictures and is one of the very popular devices particularly when the statistical facts are to be presented to a layman without any mathematical background. In this, the magnitudes of the particular phenomenon under study are presented through appropriate pictures, the number of pictures drawn or the size of the pictures being proportional to the values of the different magnitudes to be presented. Pictures are more attractive and appealing to the eye and have a lasting impression on the mind. Accordingly they are extensively used by government and private institutions for diagrammatic presentation of the data relating to a variety of social, business or economic phenomena primarily for display to the general public or common masses in fairs and exhibitions.

Remark : Pictograms have their limitations also. They are difficult and time-consuming to construct. In pictogram, each pictorial symbol represents a fixed number of units like thousands, millions or crores, etc. For instance, in Example 4.22 which displays the number of vessels in the Indian Merchant Shipping fleet, one ship symbol represents 50 vessels and it is really a problem to represent and read fractions of 50. For example, 174 vessels will be represented in pictograms by 3 ships and about a half more; 231 vessels will be represented in pictogram by 4 ships and a proportionate fraction of the 5th. This proportionate representation introduces error and is quite difficult to visualise with precision. We give below some illustrations of pictograms.

The following table gives the number of students studying in schools / colleges for different years in Indian.

STUDENTS ON ROLL AT THE SCHOOL / UNIVERSITY STAGE
(As on 31st March)

(In Million)

Stage ↓	**1960-61**	**1965-66**	**1970-71**	**1974-75**	**1975-76**
Class I to XI	44.73	66.29	76.95	87.30	89.46
University / College	0.73	1.24	2.81	2.94	3.21

Source : Ministry of Education and Social Welfare.

The data can be represented in the diagrammatic form by pictogram (pictures) as given below :

STUDENTS ON ROLL AT SCHOOL / UNIVERSITY STAGE

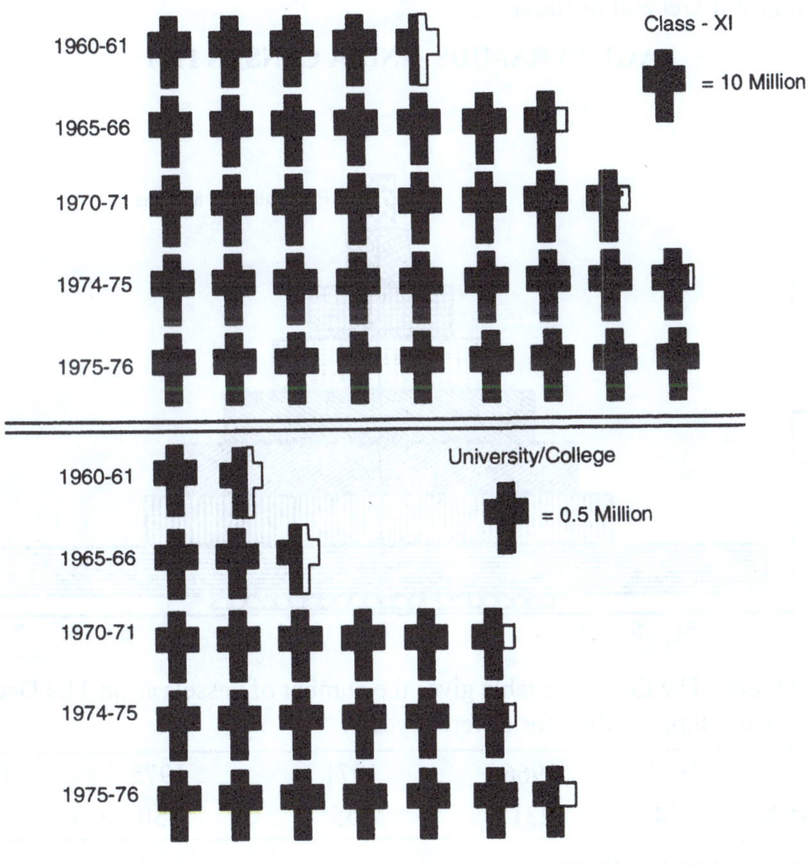

Data pertaining to the population of a country at different age groups are usually represented through pictures by means of so-called pyramids. The pyramid in exhibits the population of India at various age-groups according to 1971 census as given in the following table.

DISTRIBUTION OF POPULATION BY AGE GROUPS

(1971 Census)

Age - group	Population (in '000)
Under 1	16,519
1 - 4	63,040
5-14	1,50,776
15-24	90,569
25-34	77,010
35-44	61,186
45-54	43,416
55-64	27,202
65-74	12,880
75 and over	5,446

Source : Registrar General of India.

AGE PYRAMIDS - INDIA CENSUS 1971

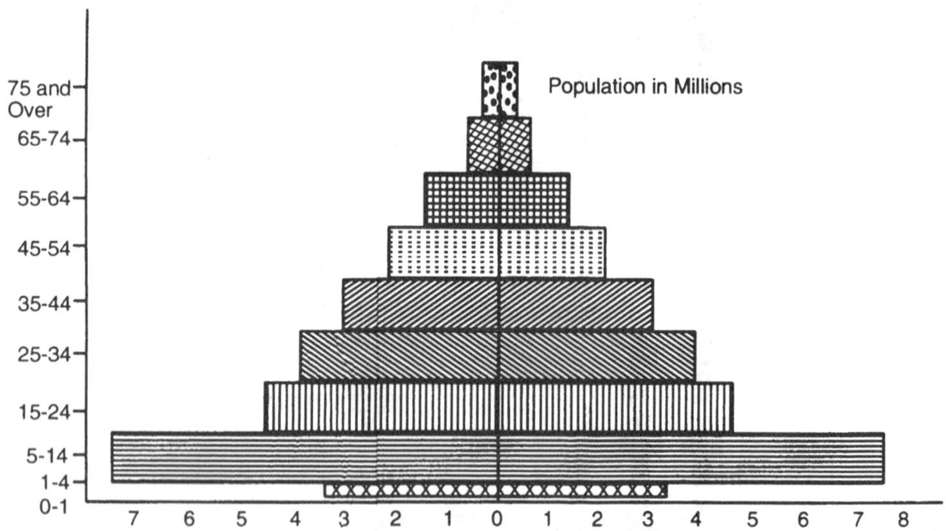

Problem : The following table gives the number of vessels as on 31st December, in Indian Merchant Shipping fleet for different years.

Year	1961	1966	1971	1975	1976
No. of Vessels	174	231	255	330	359

Represent the data by pictogram.

Solution

NUMBER OF VESSELS IN INDIAN MERCHANT SHIPPING FLEET

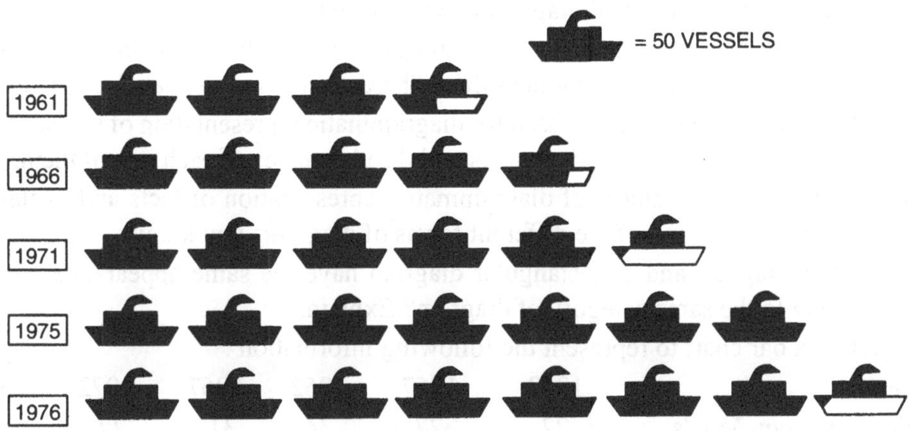

Cartograms:

In cartograms, statistical facts are presented through maps accompanied by various types of diagrammatic representation. They are specially used to depict the quantitative facts on a regional or geographical basis e.g. the population density of different states in a country or different countries in the world, or the distribution of the rainfall in different regions of a country can be shown with the help of maps or cartograms. The different regions or geographical zones are depicted on a map and the quantities or magnitudes in the regions may be shown by dots, different shades or colours etc., or by placing bars or pictograms in each region or by writing the magnitudes to be represented in the respective regions. Cartograms are simple and elementary forms of visual presentation and are easy to understand. They are generally used when the regional or geographic comparisons are to be highlighted.

Choice of a Diagram : In the previous sections we have described types of diagrams which can be used to present the given set of numerical data and also discussed briefly their relative merits and detmerits. No single diagram is suited for all practical situations. The choice of a particular diagram for visual presentation of a given set of data is not an easy one and requires great skill, intelligence and expertise. The choice will primarily depend upon the nature of the data and the object of presentation, i.e., the type of the audience to whom the diagrams are to be presented and it should be made with utmost care and caution. A wrong or injudicious selection of the diagram will distort the true characteristics of the phenomenon to be presented and might lead to very wrong and misleading interpretations.

QUESTIONS

1. "Interpretation is a fundamental component of research Process". Explain. Why so?
2. Describe the precautions that the researcher should take while interpreting his findings.
3. Give the sample table of contents of a survey based research report.
4. Assume a research topic of your choice and give the complete format of its research report.
5. Discuss the guide lines for oral presentation of a research report.

6. (a) What are the different types of diagrams which are used in Research to show the salient characteristics of group and series ? Illustrate your answer.
   (b) Discuss the usefulness if diagrammatic representation of facts.
7. "Diagrams do not add anything to the meaning of Statistics but when drawn and studied intelligently, they bring to view the salient characteristics of the data". Explain.
8. (a) State the different methods used for diagrammatic representation of statistical data and indicate briefly the advantages and disadvantages of each one of them.
   (b) Point out the usefulness of diagrammatic representation of facts and explain the construction of any of the different forms of diagrams you know.
   (c) A Bar diagram and a Rectangular diagram have the same appearance. Do they belong to the same category of diagram? Explain.
9. (a) Draw a bar chart to represent the following information :

Year	1952	1957	1962	1967	1972	1977
No. of women M.P.'s	22	27	34	31	22	19

   (b) Represent the following data with the help of a bar diagram :

Year	1970-71	1971-72	1972-73	1973-74	1974-75	1975-76	1976-77
Notes in Circulation (Rs. crores)	4,221	4,655	5,272	6,159	6,231	6,572	7,778

Scource : RBI Bulletin, March 1977.

   (c) In a recent study on causes of strikes in mills, an experimenter collected the following data.

Causes :	Economic	Personal	Political	Rivalry	Others
Occurrences (in percentages) :	58	16	10	6	10

Represent the data by bar chart.

   (d) Below are data on the number of films made in different regional and/or other languages in India in different years.

Year	1947	1951	1961	1970	1971	1972	1973
No. of films	281	229	303	396	433	414	448

Draw a bar chart to represent the above data.

10. (a) Represent the following data by a suitable diagram :

Item of Expenditure	Family A	Family B
Food	200	250
Clothing	100	200
House Rent	80	100
Fuel and Light	30	40
Education	90	210
Total	500	800

Hint. Draw Rectangular Percentage Diagram.

(b) Represent the following data by a precentage sub-divided bar-diagram:

Item of Expenditure	Family A(Income Rs. 500)	Family B (Income Rs. 300)
Food	150	150
Clothes	125	60
Education	25	50
Miscellaneous	190	70
Saving or Deficit	+10	-30

(c) Represent the data relating to the cost of construction of two tables by means of a percentage diagram.

	Table I ('00 Rs.)	Table II ('00Rs.)
Wood	5	10
Other materials	2	8
Labour	2	5
Other expenses	1	2
	10	25

(d) Draw a suitable diagram to represent the following data on livelihood patterns in India, U.S.A. and U.K.

Occupation	India	U.S.A.	U.K.
(1) Agriculture and Forestry	71%	13%	5%
(2) Manufacture and Commerce	15%	46%	55%
(3) Other Industries and Services	14%	41%	40%
Total	100%	100%	100%

11. (a) Construct a multiple bar graph to represent Imports and Exports of a country for the following years:

Year	1992-93	1993-94	1994-95	1995-96	1996-97
Imports (In billion rupees)	19	30	45	53	51
Exports (In billion rupees)	20	25	33	40	51

(b) The following data gives the number of cinemas in India.

NUMBER OF CINEMAS
(As at end of period)

Type	1960-61	1965-66	1970-71	1975-76	1976-77
Permanent	3,579	3,889	4,590	5,650	5,845
Touring	1,241	1,740	2,711	3,367	3,322
Total	4,820	5,629	7,301	9,017	9,167

Source : Ministry of Information and Broadcasting.

Represent the data by a suitable bar diagram.

(c) Draw a suitable diagram to present the following data :

	1998	1999
I Division	16	12
II Division	40	44
III Division	60	72
Failures	44	34
Total No. of candidates	160	162

12. (a) Represent the following data by data by a sub-divided bar diagram:

College	No. of Students				
	Arts	Science	Commerce	Agriculture	Total
A	1200	800	600	400	3000
B	750	500	300	450	2000

(b) The following data relate to the expenditure of the three families per month :

Items of Expenditure	Family A	Family B	Family C
Food	40	60	160
Rent	20	40	150
Clothing	20	30	100
Education	10	40	80
Litigation	5	10	30
Miscellaneous	5	20	80

Represent the data by percentage bar diagrams. Suggest a modification to this representation.

13. (a) Describe bar diagrams used in Statistics. The following table gives the daily expenditure of two families A and B Compare these figures by a suitable diagram.

Item of Expenditure	Family A (Income Rs. 500)	Family B (Income Rs. 800)
Food	140	240
Clothing	80	160
House Rent	100	120
Education	30	80
Fuel and Lighting	40	40
Miscellaneous	40	80

(b) Represent the following data by a deviation bar diagram :

Years	Income (in croers of Rs.)	Expenditure (in crores of Rs.)
1994	15	18
1995	16	17
1996	17	16
1997	18	20
1998	19	17
1999	20	18

14. (a) What do you mean by two-dimensional diagrams? Under what situations they are preferred to onedimensional diagrams?

    (b) Describe the (i) square and (ii) circle diagrams. Discuss their merits and demerits.

    (c) What do you understand by Pie diagrams. Discuss the technique of constructing such diagrams.

15. The following table gives the average approximate yield of rice in kg. per acre in three different countries. Draw square diagrams to represent the data :

Country	A	B	C
Yield in (Kg.) per acre	350	647	1120

16. (a) Draw squares or circles or represent the data given below :

Countries	Production of a Commodity (1978-79)
India	2,75,00,000
Jawa	1,55,00,000
Hawaii	83,50,000
Columbia	5,10,000

(b) Represent the following data on production of Tea, Cocoa and Coffee by means a pie diagrams.

Tea	Cocoa	Coffee	Total
3,260 tons	1,850 tons	900 tons	6.010 tons

17. (a) Point out the usefulness of diagrammatic representation of facts and explain the construction of volume and pie diagrams.

    (b) A Rupees spent on 'Khadi' is distributed as follows :

	Paise
Farmer	19
Carder and Spinner	35
Weaver	28
Washerman, Dyer and Printer	8
Administrative Agency	10
	100

Present the data in the form of a pie diagram.

(c) Draw a pie diagram for the follwoing data of Sixth Five-Year Plan Public Sector outlays :

Agriculture and Rural Development	12.9%
Irrigation, etc.	12.5%
Energy	27.2%
Industry and Minerals	27.2%
Transport, Communication, etc.	15.4%
Social Services and Others	15.9%
16.1%	

18. (a) Draw a Pie diagram to represent the distribution of a certain blood group 'O' among Gypsies, Indians and Hungarians.

	Frequency			
Blood group	Gypsies	Indians	Hungarians	Total
'O'	343	313	344	1000

(b) A ship has four compartments labelled 1, 2, 3 and 4. The space limits of 1, 2, 3 and 4 are respectively, 180,000 cubic feet, 160,000 cubic feet, 140,000 cubic feet and 120,000 cubic feet. Present the data about the different space limit in a table and draw a Pie diagram to represent the above data.

(c) Represent the following data of the distribution of income of a leading company by a suitable diagram.

	Rupees in Lakhs
Raw materials	1,689
Taxes	582
Manufacturing expenses	543
Employees	470
Other expenses	286
Depreciation	94
Dividends	75
Retained Income	51
	3,790

19. (a) Represent the following data by means of circular diagrams.

Year	No. of Employed			
	Men	Women	Children	Total
1951	1,80,000	1,10,000	70,000	3,60,000
1961	3,50,000	2,10,000	1,60,000	7,20,000

(b) Represent the following data by Pie diagram.

Items of Expenditure	Expenditure (in Rs.)	
	Family A	Family B
Food	150	120
Clothing	100	80
Rent and Education	120	80
Fuel and Electricity	80	40
Others	90	40

(c) The broad allocation of resources among the main heads of development in the Five Year Plan as compared to the draft outline is indicated as below :

Outlay ub crores of rupees 1951-56

Heads of Development	Expenditure (in Rs.)	
	Five-Year-Plan	Draft Outline
Agriculture and Community Development	361	192
Irrigation and Power	562	450
Transport and Communication	479	388
Industry	173	101
Social Services	340	254
Reforms	85	79
Miscellaneous	50	29
Total	2,050	1,493

Calculate the percentage of total outlay for each head of development according to Five Year Plan and Draft and Draft Outline and exhibit the same by Pie diagrams.

■■■

CHAPTER

11 COMPUTER APPLICATION IN RESEARCH

After going through this chapter, you will be able to know :

- Computer Introduction
- What is a Computer
- Characteristics of a Digital Computer
- Types of Computers
- Classification of Computers
- Limitations of the Computer
- Advantages of Computer
- Disadvantages of Computer
- Evolation of computers
- Hardware and Software
- Use of Binary Number System
- Network, Internet and E-Commerce
- Types of Network
- E-commerce
- Capabilities of Computer in the filed of Research

The use of computer is becoming a very common in analysis of data. Research students must use all the methods available for interpretation of data. Therefore, it is necessary that, researcher must have knowledge as to how computers can be used in research activity.

In this chapter the information in relation to computer application in research has been explain. So that, proper application of the computers can be made. This aspect has been introduce in the Simplest form. Like what is computer, how does in function and how data can be analysised in computer.

Computer Introduction

The discovery of numbers, which is considered as the greatest achievement of modern knowledge, led to the development of all kinds of modern calculating machines. Today nobody can imagine a world without numbers. It took thousands of years to get the present system of computer knowledge. This development passed through several stages. The development of first calculating device named Abacus dates back to 3000 BC. From Abacus to the Micro computer, the counting system have undergone a tremendous change. The efforts are still continuing to make micro computer more and more useful. Thus history of computers is such a story, which is illustrated with the names of number of architects whose inventions and discoveries perhaps did not bear any direct significant at that time, to the development of modern age computers.

The most early methods of simple calculation were done with the help of counting pebbles and twigs or cave making or tally bars as used in statistics today, and even the use of one's own fingers and toes. This in fact is how the decimal system of counting with a base of 10 devolved because we have 10 fingers and 10 toes.

The stone age man used pebbles for counting his cattle. When the sheep left its fold the herdsman put one stone in his pocket and continued doing so till all the sheep left the fold. When the sheep returned to the fold, he repeated the above procedure in reverse order. In this way stone age herdsman laid the foundation for a very important concept of modern mathematics. One to one correspondence. These stones were later replaced by sticks, scratches, symbols on stones and knots on a rope. Latter these scratched and symbols were replaced by the present number system. The discovery of zero by Indian mathematicians laid the foundation stone for present numeric system.

A computer is an electronic calculating machine that receives data/instructions and processes them to produce desired output required by user.

It is mainly used to store the data, so it could be easily retrieved, whenever needed by user. It is also known as a high speed calculating device for performing millions of calculations within nano as pico seconds with great accuracy.

What is a Computer

A computer is a device that accepts information and manipulates it for some result. Computers can be confusing, and it takes a little time before you gain confidence in using and understanding them. Just like a new TV or VCR, computers can be a bit cumbersome to figure out, but once you achieve a certain level of knowledge, these machines can become powerful allies.

Following represents the block diagram of computer

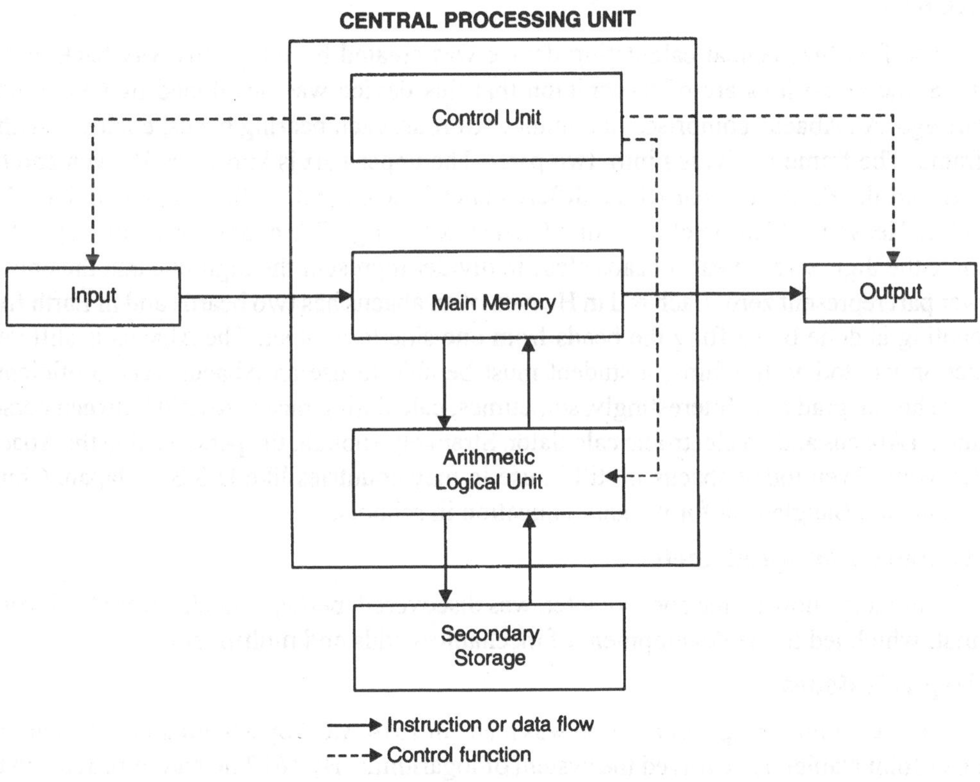

A computer collects processes, stores and outputs information as per the following order:

Input

An input device help us to communicate with a computer. You can use input devices to enter information and issue commands. A keyboard, mouse and joystick are input devices.

Process

The Central Processing Unit (CPU) is the main unit in a computer. The CPU processes instructions, performs calculations and manages the flow of information through a computer system. The CPU communicates with input, output and storage devices to perform tasks.

Output

An output device lets a computer communicate with us. These devices display information on a screen, create printed copies or generate sound. Monitor, printers and speakers are output devices.

Storage

A storage device holds information. The computer uses information stored on these devices to perform tasks. The hard drive, the tape drive, the floppy disk and the CD Rom drive are storage devices.

Foundation Stone for present numeric system

1. Abacus

The first mechanical calculating device was created by Egyptians way back in 450 B.C. Some authorities are of the opinion that this device was developed by Chine 3000 years ago. An Abacus comprises of a number of rods, each bearing beads, enclosed within a frame. The frame is divided into two parts. The upper part is known as Heaven and the lower part the Earth. The bar which differentiates Heaven and Earth is called Divider. The beads in Heaven, which are close to divider, denote the digit five and beads in raised position denote the digit zero. In earth beads close to divider represent the digit one and the ones in lower part represent zero. Each rod in Heaven of an abacus has two beards and in Earth five. Counting is done by shifting the beads from one side to another. The Abacus is still very much in use today. In China, a student must be able to use an Abacus very proficiently before he can graduate. Interestingly, sometimes, calculating races are held between person using an Abacus and on electronic calculator. Strangely enough, the person using the Abacus often wins. Even today abacus is still in use in may countries like U.S.S.R, Japan, China, Pakistan and Bangladesh for primary education in schools.

2. Odometer or Speedometer

Odometer, now called speedometer, was discovered, perhaps much before the birth of Christ, which led to the development of mechanical adds and multipliers.

3. Napier's Bones

The next pioneering contribution was made in 1614 A.D., by a Scottish mathematician named John Napier. He evolved the system of logarithms. By 1617 he had improved on his device which could perform multiplication and division as well. This device was know as

Napier's Bones and consisted of a set of 11 rods, which were further divided into 9 diagonal numbered parts. By manipulating the rods it was possible to perform calculations within 2 numbers.

4. Slide Rule

The slide rule was invented just few years later (1620) by William oughtered. It has two marked rulers. One of which could be slide over the other and once properly aligned, would give the user the required product or quotient. This is an apt example of an' analog' device in which notations could be taken as continuous readings. This is apposed to a digital device which operates one step at a time and works with distinct Values. The essential parts of slide rule are :

1. The body of the rule which carries scales. This part is stationary.
2. The slide carries scales and is movable, and
3. The cursor or hairline indicator.

A simple slide rule consists of four scales called A, B, C and D. Two of these scales are on the body of the rule and the other two are on the slide. We can perform multiplication, division, squares, square roots, proportions, percentages and linear equations with it.

5. First Programmer

Lord Byron's daughter, is famous as the First Programmer, for having devised a suitable use of the Binary Number system for programmes and data to be fed into the Analytical computer.

6. Boolean Algebra

George Boole actually applied the selection of alternative choices and its link with computer development. His theoretical development for representing and manipulating logical evaluations in terms of true or false, formed the basis of 'Computer Logic' and in being applied to the algebraic system came to be known as Boolean Algebra.

7. Magnetic Media

Major development were taking place in the areas of recording devices by Mr. Valdemer Poulson. He devised the method of tapes and drums coated with thin films of magnetic materials. This is regarded as the forerunner of the present day magnetic media-cassettes, tapes, video tapes, etc.

8. First Automatic Electro-Mechanical Computer

In 1930 at Harvard University, *Howard Aiken* and *Grace Hopper* got together and arranged for the IBM to build them a general purpose computer. They called it the Mark I. This machine was very reliable and is regarded by some as finished engine. Now this machine was really, something 50 feet long, it had 500 miles of wires and thousands of relays. If a relay broke it was really hard to find it and fix it.

9. Punched Cards

A big step forward was taken in the development of information processing in 1890 census of U.S.A. The U.S.A. Census Department was looking for a quick way of processing the census information be put on punched cards, with a special code devised for representing all numbers and letters. For this purpose he designed a machine to punch the collected data on the cards. For sorting and counting these punched cards he designed a tabulating machine.

The working of the Tabulating machine was each card was put through this machine that had little rods that touched the card. Whenever there was a hole, a rod would go through the hole. This would allow electricity to flow through the rod. The number on the card could then be counted automatically. With the help of this machine the results of the USA census 1890 were ready in just two and half years.

Characteristics of a Digital Computer

Computer System = Hardware + Software

Hardware = Internal Devices + Peripheral Devices

Hardware is the term associated with Computer Architecture.

All the parts of the computer are hence referred to as computer hardware.

The chip level circuitry is well designed inside the processing box with the two input/output units of the computer.

According to the block diagram of the computer, the computer is devided into four main parts as

Input______ Processing ______ Output______ Storage

The Input Unit fetches the data to be processed.

The processing unit converts the data into meaningful information.

The Output unit displays the processed information to the user which may be in the form of the display unit or the storage unit of the computer.

CPU or the Central Processing Unit

The main component to make a computer operate is the computer chip or microprocessor. This is reffered to as the Central Processing Unit (CPU) and is housed in the computer case. Together, they are also called the CPU.

The CPU acts like a traffic cop in a busy intersection. It instructs the computer programs to process the data.

Power Switch

The power switch is located somewhere on the CPU, usually in the front or on the side. You press this to start your computer. But **do not press the power switch to turn off your computer** unless nothing else works.

Motherboard

All of the computer's components are hooked up to a motherboard, which is the main base for the computer's circuitry and components.

Ram

Memory is called RAM, which stands for **Random Access Memory.** RAM is the actual grouping of computer microchips on your motherboard.

Megahertz

Your computer speed is called megahertz, after Heinrich Hertz, a German physicist. It is defined as the millions of cycles that the electric circuitry goes through while operating your computer.

Elements of Computer are :

1. Hardware
2. Software
3. User (People)
4. Data/Information

Characteristic of Computer

The characteristics of the computer are as follows,

(1) Speed

Computer is a fast calculating device. It calculates millions of instructions in seconds. This characteristic is called speed of computer. It is this characteristics, which is responsible for the wide usage of the computer. The computer with its speed has the upper hand over the humans in solving calculations complex. The speed of the computers is measured in microseconds (10^{-6}), nanoseconds (10^{-9}) and even Pico seconds (10^{-12}). The calculation which will be performmed by human in days or weeks, is done by the computer in seconds. Thus with its speed the computer has become a great boon for ever-growing technology.

(2) Accuracy

It is a human tendency to commit mistakes during long hour working, which may cause lot of trouble. The computer on the other hand gives accurate results. The degree of accuracy of a computer depends on it's design, but it is sure that computer never commits any mistake. The errors in the program or wrong feeding of data are only caused by the human and are normally called as human errors. As far as computer is concerned, it is well known for its accuracy and therefore has great commercial value, thus it can solve many calculations and execute other instructions without any mistake with high precision and great degree of accuracy. The most important thing with respect to computer accuracy is that its accuracy is consistent and reliable.

(3) Deligence

The human get tired after working for a specific period of time and can cause erros. This is where the computer scores over the human. The computer is a machine, therefore it is free from the feeling like tiredness and hence can work forever, which is an advantage of computer. Even with lots of work, accuracy remains consistent. This characteristic of the computer is called Diligence.

(4) Versatility

One of the important factors of the computer is that it is not limited to a particular kind of work. It alone can perform various types of work single handedly with great speed and accuracy. It is one of the important features of the computer. If it deals with solving complex calculations at one time, even at other time it may busy in chatting on Internet or in taking print outs of a particular job. It performs many tasks simultaneously. This is the versatility of the computer.

(5) Storing capability

The human find it difficult to remember any information for long period, and when it is recalled it is not adequate or not in sequence. So remembering all data for more days is almost impossible for a normal human being. The computer will store even large volumes of data in its memory for long periods and when the data is retrieved it will be presented in the same sequence, as it was stored. Nowadays many devices like magnetic types, magnetic disks or other storing devices are used with the computer for storing large volumes of data. This feature of the computer is extremely helpful to human and assists human in their work for making decision.

No Decision making :

Although, the computer can remember large amount of data, it can perform all the complex calculations within seconds, may perform lots of work efficiently, being versatile and possess all other qualities, but it cannot make a decision on its own. It only performs as instructed by the user. Nowadays Scientists are working on Artifical Intelligence Technology, so that computer can take decision on its own to some extent.

In brief, Computer processes the instructions very fast with accuracy, store large volume of data more efficiently than human being. But the computer does not have commonsense (i.e. it can not take any decision itself without human help) which is possesed by human beings.

Types of Computers

Computers can be classified by their size and power

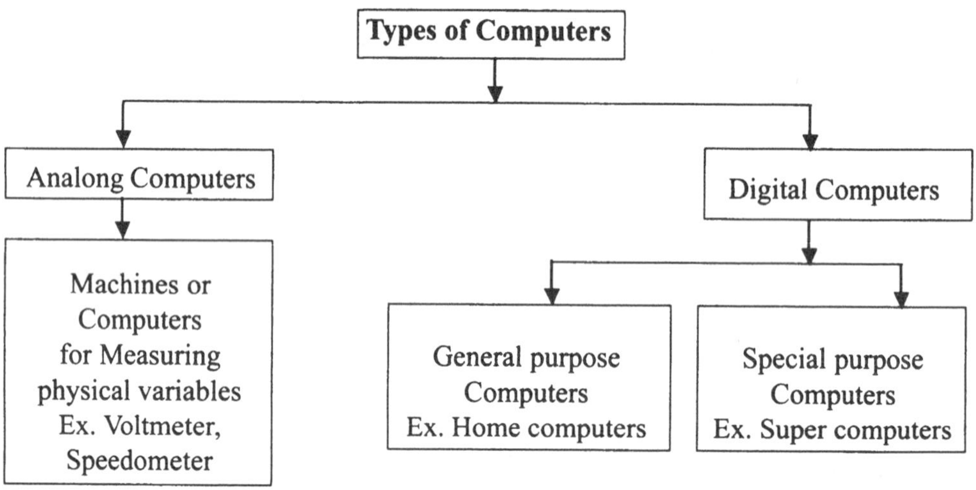

There are normal two types of computers :

1. Analog Computers
2. Digital Computers

(1) Analog Computers :

These types of computers are specially built for a specific purpose. These computers usually measure physical variables or physical parameters such as pressure, temperature, speed, voltage, currents etc. Thus, these computers are used for measureing purposes. Therefore they are not very precise, consistent, have very limited applications and lacks versatility. For ex. Speedometer of cars, voltmeters and ammeters used in laboratories.

(2) Digital Computers :

The term Computer used in our day-to-day life is referred as a digital computer. These computers work with discrete type of values. In digital watch the time display does not vary continuously, but altered into a new digital format. They are more useful in solving expressions, complex and statical problems. These types of computer are normally made for general purpose having good speed and store result. The storage, calculations and manipulation are done in bit format i.e. in the form of 0 (Off/Low) and 1 (ON/High).

Digital computers are again classified into,

- General Purpose Digital Computers
- Special Purpose Computers

General Purpose Digital Computers

The digital computers, which can be used for all general purposes like in business, in solving several mathematical expressions without any difficulties or complexities. In short they are the computers, which fulfill our general objectives and are called as general purpose digital computers. Ex. Home computers, PC's.

Special Purpose Computers

These types of computers are made for some special objectives or some special purposes. They possess high speed and are very costly. They are too big and costly for our general purpose needs. Therefore there are not normally found in homes. They are used by the organisations, to work out specific task. They perrform the computations, which are far beyond the reach of general-purpose computers.
Ex : Super computers working at missile launching stations.

Hybrid Computer

It combines the best of features of Analog and Digital computer. These computers are useful for solving special problems in which input data is in the form of analog signal, which may require to convert into digits and processed by the computer.

Classification of Computers

Computers are classified into four types,

1. Micro computers
2. Mini computers/Desktop computer
3. Main frame computers
4. Super computers

The above classification was done according to the computer architecture, spped, protability, cost and other differences.

(1) Micro Computers

The micro computers are the computers in which microprocessor is the central processing unit of the system. It is a small-integrated chip, which performs all the functions of the system like logical and arithmetic calculations and all other process necessary for the system.
Ex. : Home computers and PC's in organizations.

Purpose : Small business units and office automation system normally need them. They can also be used in the other areas like computer literacy, in playing games, programming, Telecommunications, Data base management, According and Word processing. It consists of a terminal i.e. a keyboard and a CPU connected with the monitor.

(2) Mini computers

Mini computers in 1960s and 1970s were used for many computing tasks with good speed but with less power. Compared to main frame they were portable and less expensive.

The mini computers were soon started to be used as a time sharing or multi-user system and later there was great increase in speed storage and other characteristics with reducing cost. Multiuser system used by medium sized organization.

(3) Main frame computers

The high-speed computers (Compared to Micro and Mini computer) used in financial organizations, which has the speed of millions of instructions per second (MIPS) are called as Main frame computers. One of such computer IBM 3090/600 is rated at 102 MIPS. Nowadays the important role of the main frame computers is to support large database. In business and government organizations large data transactions take place therefore it is necessary to have main frame computers in such organizations. The main frame computer is very costly compared to mini and micro computer.

Limitations of the Computer

1. Computer lacks thinking power, they perform tasks instructed by user / programmer.
2. Even for routine works, they have to be instructed, since they can not learn unlike humans.

Computer in Business

Computer is used for storing information and business related records. This information helps in taking useful decisions of current and future prospects for company, by management. They are used for inventory, sales analysis, credit analysis, cost accounting, operating ratios, manufacturing schedules etc. They also help in retrieving records, if required by the user at anytime. Nowadays, the success of the company relies on the amount of computerized work carried out by that company, since it is reliable source.

For taking a better decision an information must be,

- Accurate
- Action-oriented
- Commensurate
- Concise
- Timely available

It is helpful in following areas,

1. Sales and Marketing

Computer is greatly used in this particular area, since it can store all the records of current sales, estimate future trends and the future's changing demands. With the help of computer, Marketing Executive could plan the marketing strategy and identify the best marker for their product.

2. Accounting

In accounting, Computers help in taking printouts of bills, profit-loss statements, tax reports and other necessary documents. It can evaluate production cost and compute hourly costs for individual machines or entire departments thus, enabling the company in taking decision.

3. Personnel Management

It is used in maintaining information and records of Employees capabilities for future use. It can list employee by department, salary, schedule or seniority, prepare overtime allotments, which could help the management in taking decisions in promotions of their employees and help the company in a best possible manner.

4. Production, Planning and Control

It is used in planning the production of the particular company. To prepare their work schedule for various employees. Thus could best meet the needs for the company's future demands.

Automation

Once a programme is in computers memory, CPU executes the individual instructions with or without little intervention of humans until completion of process or job. This is called as Automation. Ex. Printing of Electricity bills.

Advantages of Computer

1. It increases the productivity in industries.
2. It attains a consistently high quality of product.
3. It improves the service.
4. It extends and existing manual system.
5. It increases speed and efficiency of operation.
6. It handles lots of work without increasing staff or accommodation.
7. It provides manament with information more quickly.
8. It keeps current and future costs of work, down.
9. It eliminates lots of work.
10. Because of many tasks involved in collection and handling of more and more information, many jobs have been recreated in many other areas.
11. In industries, the computer takes over many routine, semiskilled jobs, some of which are hazardous to human.
12. The use of computer technology will create new products and services, and hence new jobs.

Disadvantages of Computer

Automation in many areas has significantly changed the work pattern. It has caused threats to some jobs and seems to bring more changes in future. Jobs are likely to be squeezed. Activities like Word processing, Data Base Management Systems and Electronic may merge may together in future, bringing a high degree of automation to the offices, other than this Automation also includes the growth of industrial robots, which will carry out work without human intervention.

Industrial robots are programmable device designed to manipulate tools, it can transport parts and perform manufacturing tasks. A computer programme can directly control its movement and actions. Similar is the principle of automated car where Robots carry out all the work.

Intelligent Robots are anticipated in the future, equipped with some measure of visual perceptions and they can take small decisions by themselves.

Evolution of computers

The greatest milestone of the achievements made by mankind has been the discovery of the numbers. In early days man used stones and pebbles for counting. With the invention of zero by the Indians, the number system developed with a new pace.

The later growth of trade and commerce gave way to further advancements on the field of calculation. The following important developments enroute the modern day computer which is no more an instrument but a household name.

The following are some of the important developments which took place contributing to the present day world of desk top machines.

Generations of Computer Technology

The First Generation (1940-1952)

The first generation computers were large in size, unreliable and required a constant maintenance. They were based on the technology of electronic valves (Vacuum Tubes). They were based upon Von Neumann's design principles. The computers belonging to this period were :

(a) EDSAC (Electroinc Delay Storage Automatic Computer) : it was the first computer to operate on the stored program concept. It was built at Cambridge University by M.V. Wilkins and was first used in May 1949. This machine utilised mercury acoustic delayed lines as storage. The cycling process of data represented by a set of impulses, spaced periodically, was kept circulating through the mercury until needed.

(b) EDVAC (Electronic Discrete variable Automatic Computer) : It was designed at the University of Pennsylvania for the United State Army in 1950. It used the sonic delay properties of mercury for storage. it had th distinction of being the first computer with commands and programs stored in it.

(c) ACE (Automatic Computer Engine) : It was built at the National Physics laboratory in 1951 and provided pace to the development of the Computers.

(d) LEO (Lyons Electonic Office) : It was developed by M.V. Wikes in 1951 and was the first commercial computer.

(e) UNIVAC (Universal Automatic Computer) : It was produced by UNIVAC (Universal Accounting Company) in the year 1951. It was established by Eckert and Mauchly. It was the first machine to use magnetic-tapes as an input/output medium. It had the capability to running 24 hours a day. It was reliable and had the ability to handling both numbers and large amount of information.

The Second Generation (1952-1964)

The second generation of computers used the technology of transistors. The transistors were invented by a team of scientists headed by William Shockley at the Bell Laboratories in America. They are made out of materials known as semi conductors (Germanium and Silicon mainly), by doping layers of crystal with impurities which alter the electrical properties of each layer. They performed similar properties like Valves but were cheaper, smaller and relaiable. They were bistable in nature and brought in the probability of the two state devices i.e. the concept of Binary numbers. The computers belonging to the second generation of the computers had inbuilt error detecting devices. The storage capacity was increased due to the incorporation of magnetic core memories. There was a unique development of High level Languages and assembly languages during this period which brought the computers more familiar to the common man. The first igh Level Langauge FORTAIN (Formula Translation) was released in the year 1957, by the international Business machines. The computers belonging to this generation were, IBM 7000, LEO MARK II, NCR (National Cas Register)-304, ATIS, CDC, RCA, UNIVAC 1108, etc.

The Third Generation (1964-1972)

The computers belonging to this generation used Integrated (Transistor) Circuits (IC'S) instead of Transistors. The development of ICs allowed future prospects of the development as it was more relaible and reduced the size of the computers. Instead of having one transistor of its own, several transistors were integrated and sealed in an integrated circuit called IC. During this period SSI (small scale integration circuits) and MSI (the medium scale integration circuits) were used. The internal storage capacity was further increased, there was a marked increase in the processing speeds, the access time was reduced to 100 nano seconds and there was extensive storage and input/output capabilities. The concept of Time sharing and Batch processing was also developed during this time along with the invention of high level languages like BASIC and COBAL. Some of the third generation computers were IBM 360/370 series, ICL - 1900/2900/2903, PDP-11/8/45, and Honeywell-6000 series.

The Fourth Generation (1972-1985)

This period marched with an active research and development. LSI (Large scale integration circuits) and VLSI (Very large scale integration circuits) were used in the fourth generation brand of computers. As a result the microprocessors came into existence. The computers using this technology became to be known as micro computers. With the march of time during this period only, the developing pace brought the computers having a processing speed ten times faster, memory capacity four times greater and the size 1/10th times smaller of the earlier IBM or the last generation computers the storage capacity of the computers were increaed to almost 1 MB and greater. The accessing time got further reduced to 1 nano second. Computers belonging to this generation were - STAR 1000, DEC 10, PUP11, APPLE1/11, and DCM series.

The Fifth Generation (1985 Onwards)

The fifth generation of the computers have had the union of software and hardware to produce the concept of artificial intelligence. These computers mark major chnage from the earlier generations of computers and are equipped with reasoning and decision making capability like human beings. They are termed KIPS (Knowledge Information Processing Systems). They are to use fibre optics, bubble memory and voice recognition. The great super computers belong to this generation of computers.

The concept of Artificial Intelligence has associativity with the fifth generation machines. The artificial intelligence is a computer software that allows a computer to perform tasks which are considered "intelligent" if done by an individual. These tasks include giving expert advice, understanding natural language, speaking intelligently and recognizing complex patterns such as handwriting. The three most important types of Artificial Intelligence Programs today are Expert System, Natural Language Programs and Neural Networks.

The major applications out of this concpet highlight as :

(a) Embedded Artificial Intelligence within a database software.

(b) Robotics, which incorporate an insect like electronic creatures with a simple intelligence.

(c) Robot Eyes or Intelligent Vision having the feature of identifying various components on a complex circuit board.

(d) CASE - Computer Aided Software Engineering, which helps in designing various utility software.

(e) CAI - Computer Aided Instruction, which acts like a tutorial software for automatically adjusting the knowledge of an individual.

With pace of time and tide there is an accelerated progress to the branch of artificial intelligence. The software experts have come up with a novel concept of virtual reality incorporating A.I. The Virtual Reality or VR software, along with special goggles and gloves, allows individuals to seemingly enter an artificial world that exists only within the computer, and even move and touch objects that appear in the 3-D view provided by the goggles.

This new generation offers super high-speed Pentium PC's which have become a common platform of choice for development of A.I. Software.

Expert Systems

An Expert Ssytem is a computer that can solve real world problems required logic, decision making and knowledge processing. Expert Ssytems can also categorize, consult, analyze and diagnose. These software tools are useful in areas that formerly required the human expertise. They use deductive reasoning to solve problems that are often unstructured, and would be impossible to approach with conventional procedural computer techniques. It acts an application and guides the performance of tasks which usually require experience and specialized knowledge and it is hence being widely used in decision support. The data and decision rules which represent the expertise is known as the knowledge base and the area of human intellectual endeavor to be captured in an expert system is called the domain. One might think that when consulting an Expert System, one may ask the Expert Ssytem a question, and receive a solution. Logically, the vast majority of Expert Ssytems in use today ask the individual the questions and arive at a soution based on his responses using the domain. Teh following example follows an encounter to the concept :

Question 1 : Do you have a cough.

Your Answer : Yes

Question : Do you smoke.

Your Answer : No.

Solution : You are suffering from cold.

The above solution is based on the knowledge base of the software and hence requires a lot of expertise of dictionary.

The shell of an expert system is an expert system minus its knowledge base and the domain. Some of the common expert systems can be found in the area of medicine and computer configuration, and they surpass in totality the human expertise. The remarkabke example of DEEP THOUGHT the computer which defeated the Grand Master Chess players has the conduct of the unique application of the Expert System interface.

Natural Language

As another important branch of Artificial Intelligence, the Natural Language is a software which focuses on getting the computers to understand the spoken or typed English words. The ability of the machine to conduct of understanding of the spoken words and let the machine to speak out the typed words has become possible with the application of the feature of Artificial Intelligence. There are software which can speak out any text coming on the screen and can type the slowly spoken words. There are also novel features as content to automatically translate from one language to another.

Neural Networks

It is a computer application that attempts to mimic the neurophysiology of the human brain in the sense that the network learns to find patterns in data from a representative data sample. It acts as a biological model of artificial Neurons, connected to each other by Axions. Each Neuron can have many inputs, but only one output. As a Neuron gets energized by input, it fires, sending energy along axions to other nearby Neurons. If another Neuron receives enough energy it will also fire, propogating the excitation to others. There are hundreds or even thousands of such Nuerons, arranged in layers, and all together they form a Neural Network, capable of learning form experience through a self correcting technique called back-propagation. Presently, the Neural Network is the most exciting area in artificial Intelligence, and the machines incorporated with this capacity routinely perform outstanding feats of pattern recognition such as reading scribbled handwriting at high speed with incredible accuracy, learning to identify male versus female faces, and evenlearning to speak all most without human intervention.

Accordingly, neural networks are especially useful for analytical problems involving :

Lots of example data, where it is difficult to specify a prarmetric model for the data.

Multiple relationships within the data that are not fully understood with many probable specifications.

Containing fuzzy data, with stable and deeply hidden patterns.

Generation of predictions involving complicated explanations.

The common applications of Neural Network involve credit risk assessment, market segmentation and sales prediction.

Hardware and Software

A Program is defined as the set of instructions given to the computer for the execution of a particular task. A set of such programs is termed as a SOFTWARE.

the term software describes all the written procedures and rules that controls the computer operation. The instructions are given to the computer in the written program which, when fed into the computer, makes the computer behave the user wants it to.

In other terms, software is a set of electronic instructions that tell a computer what to do. You cannot see or touch software, but you can see and touch the packaging the software comes in they are classified into two categories namely, system software and application software.

System Software

It is supplied by the manufacturer and is specific to the machine hardware. It has the following categories featuring the operation of the matching, by the user.

(i) Assemblers/ Compilers/ Interpreters

(ii) Operating system and

(iii) Utilities

The assemblers/ compilers / interpreters are the set of programs that convert the list of instructions written in high-level langauge (English like language) into the machine level language (o's and 1's) or vice versa. The program which is written in ahigh level language is termed as a SOURCE PROGRAM and a program which is written/converted into machine language is termed as an OBJECT PROGRAM.

An Assembler is a program which is used to convert a program written in Assembly Language into Machine Language and vice-versa.

A Compiler is a program which translates a set of instructions written in Assembly language into Machine Language and vice-versa as a whole.

An Interpreter is a program which translates a set of instructions written in High Level Language into Machine Language and vice-versa line by line.

The Operating system is the link between the hardware and the software. it is a set of programs that are loaded onto the computer's memory (RAM) and makes the computer functional. It manages the files and runs the program. Operating system software controls the overall activity of a computer. Highly Advanced computers use Windows 2000 as an operating system software.

The Utilities are the accessories along with the operating system. They facilitate the user for the smooth operation of the machine and conduct added facilities. They are designed to save computers's time from repetitive and unnecessary programming. For the execution of these utilities. The presence of the utility file concerned is a must. e.g. To format a floppy disc, the presence of the utility file FORMAT.COM is must. The Utility Program or Utility File is an independent file required to assist the computer and is a part of the system software.

Application Software

It is the set of those programs (the list of instructions) which are used to perform some application with the aid of computers. Application software lets you accomplish specific tasks. Popular application software includes Microsoft Office XP

They are developed for actual usage. Some of the software programs used for application are :

(a) A word processor or an editor (line and screen), to type the text with dacilities of modification, deletion, and editing.

(b) A data base program to structure data into related fields and to check the validity of input.

(c) A report generator to print formatted reports from a given file or data base. etc.

The examples follow the most commonly used application softwares are Windows, MS-Word, MS-Excel etc.

Use of Binary Number System

The only data that a computer can understand is on and off. But, the millions of combinations of those two commonds given in series are what make a computer work. the memory chips inside your computer are divided into thousands of tiny compartments called bits. Each bit has an electronic switch or gate. On means the gate is open and letting electricity go through. The computer reads on or open switches as a number 1. Closed gates are off because the electricity is blocked and cannot get through. The computer read off bits as 0. It is by grouping these bits together to form a series of 1/0 commands, that data is formed.

Eight bits are grouped together to form a byte. In this group of eight, there are 256 possible combinations of 1/0. The grouping of 1/0 within a byte is called **Binary Code.**

Byte

One byte is one cahracter. A character can be a number, letter or symbol. One byte consists of eight bits (binary digits). A bit is the smallest unit of information a computer can process.

Megabyte (MB)

One megabyte is 1,048,576 characters. This is approximately equal to one novel.

Kilobyte (K)

One kilobyte is 1,024 characters. This is approximately equal to one page of double-spaced text.

Gigabyte (GB)

One gigabyte is 1,073,741,824 characters. this is approximately equal to one thousand novels.

Nibble

A combination of four bytes make a nibble.

Also : One KB means 2^{10} bytes
One MB means 2^{10} KB
One GB means 2^{10} MB
One TB means 2^{10} GB

Uses of IT

The numerous advantages of IT are pretty obvious. For example, by simple accessing a website, information can be obtained at the click of mouse, saving the student much precious time walking to the library to reserve or refer to a book. IT nowadays also provides easy access to materials through, say, e-books and e-journals. Online discussion forums between lecturers and students are also possible.

IT is able to perform many functions that the lecturer is unable to fulfill. For example, it can be used to demonstrate animated examples of certain machinery (in the case of engineering), without which the lecturer would need to arrange for the students can still communicate with him or her, as well as with each other through IT facilities. This enables eduction to free itself from the confines of the classroom.

Nowadays, there are online courses made available to interested parties from all walks of life, eliminating the need for students to travel to a specific location for lectures. They can even be in other countries. For instance, in Europe such courses attract students from many other European countries. In the US, Universities are trying to put all their courses online so that students from all over the world can subscribe to them under a distance-learning scheme. The students however may need to go to the Universities once in a while when it is really essential ot the lecturers may need to go overseas to teach their foreign students occassionally. Such online courses are also ideal foe very large classes that are too huge to fit into a lecture theatre. Students can access online lectures at their own convenience and larn at their own pace. Moreover, it allows for the inclusions of additional material and animated demos, and may end up more interesting for the students. Even remote-controlled practical laboratory sessions are possible with IT and these have been conducted in some advance countries.

With IT, lectures can be repeated in exactness, and teachers ans students can easily review and evaluate such and such a lecture at will. PowerPoint presentations are permanent records that are convenient to edit and send via the Internet. Video conferencing saves money and time in travelling and globetrotting on the part of the participants.

Limitations of IT

Despite the fact that PowerPoint presentations are extremely popular and attractive, many top universities in the US discourage their use for teaching and learning. One of the reasons is that the lecturer's train of thought could be retrained by the prepared presentation, hindering flexibility and on-the-spot reaction to changes in corcumstances. Students have often complained that such presentations are too fast for them to follow and assimilate. A certain amount of personal touch and sense of 'live performance' is also lost. Lecturers themselves admit that by flashing facts and figures using PowerPoint presentations actually squeezes too much of the syllabus into a short space of time. Using transparencies on an overhead projector has a more human touch and students can follow the lecturer as he changes transparencies. Moreover, PowerPoint presentations take ages to install and are practically useless during an equipment or power failure, requiring a backup on transparencies in any case.

Although websites are generally powerful tools for gathering information, not all websites are well designed or user-friendly. Those that are complicated end up wasting the surfer's precious time. Students sometimes complain that such and such a website is more confusing than educational. As for web casting of lectures, the inpression that students get when using these is that they feel that they are communicating with a machine rather than with a person. It has been found that distance-learning students from other countries or other universities accessing online courses tend to drop out more frequently than those within the same university for the simple reason that they just do not feel part of the learning community.

As a matter of fact, all the IT programs are limited in scope, and therefore are inflexible beyond their capacity. Judicious use of IT has its advantages. Still, it is up to man, who created IT, to know how far it can be harnessed for the purpose of teaching and larning.

Network, Internet and E-Commerce

Networking

A computer network is an interconnection of various computer systems located at the same or different places by cables or telephone or satellite and to share information and Resource (Software, Peripheral Devices, Printers etc.)

Communication device : A communication device is used to connect one computer to another. among the many kinds of communication devices, the most commone are :-

Modem (Sodulator / Demodulator) : Which enable computer to communicate through telephone lines. Thus modem is useful to convert digital signal to analog signal and vice-versa.

Network Interface Card (NICs) Which let users to connect a group of computer to share data and devices.

Types of Network :

1. Local Area Network (LAN)
2. Wide Area Network (WAN)

Local Area Network

LAN is used to interconnect computers in a single room within building or buildings. LANs normally transmit data in a digital for with typical transmission speeds between 0.1 to 100 megabytes per seconds. These speeds can be achieved using parallel transmission. Thus transmission spped is very high. the attached computers may be of different types and size and performing a variety of functions such as data processing, word processing or electronic mail etc. The cost of transmitting data in a LAN is negligible, since the transmission medium is usually owned by the user organisation. Fewer data transmission errors occur in case of a LAN.

Wide Area Network

WAN is a digital communication system, which interconnects different sites, computer installations and user terminals and may also enable LANs to communicate with each other. This type of communication network may be developed to operate wide or world wide. Transmission cost of data is high because medium used are normally public systems such as telephone lines, microwave etc. a WAN is also known as long hard network. In a WAN, there may not be a direct physical connection between various computers. Data transmission speed is normally of the order of 1800 to 9600 bits per second. A common example of WAN is Natioanl Telephone System, Cellular Mobile Phones etc.

Difference between LAN and WAN Topology :

LAN	WAN
1. Interconnect the computer terminals within buildings.	1. Interconnect the computer terminals of wider graphical distance.
2. Transmission speed of data os very high. Typical transmission speed in 0.1 to 100 mega byte per second.	2. Transmission speed of data is slow and speed is 1800 to 9600 bite per second.
3. Cost of transmit data is negligible.	3. Transmission cost of data is very high.
4. Terminals and peripherals devices are usually connected with cables and wires.	4. May not be direct physical connection between various computers.
5. Fewer data transmission error.	5. More data transmission error.

Topology :

An inter-connected set of computers is often referred as a computer network and the structure of the network is referred to as a topology. The topology determine the complexity, cost of Network and Strategy for physically expanding Network.

Several types of topologies are as follows :

1. **Ring Topology (closed loop) :** In a ring topology, information is passed from one node to the other in series. Information is divided into packets for transmission between nodes and each packet contains the address of the node that is to receive the information. Its major disadvantage is that th entire network goes down if any one of the nodes fails. It is also called circular topology.
2. **Bus Topology (multi-drop topology) :** In the multi-drop structure, nodes share a single link or communication channel somewhat similar to the multiplexed microcomputer bus sturcture. Each node has a unique address if one node is down, the network is still available for communications between the other nodes. If the link is broken, then parts of the network wll be seperated.
3. **Star Topology :** In these network the central node is often the master. Each of the other nodes are joined to the master via separate links. This network can be fairly effective, since all communications must go through the central nod, if the central node over goes down, then the complete network wll be down.

Apart from this, new variated technologies were created depending on these type of topologies.

1. Mesh / Graph / Completely network topology
2. Tree network topology

1. Mesh network topology : Connecting some or all the computers to one another, i.w. the ring network topology was modified, so that they can communicate with each other. this network looks like Mesh, hence this type of topology was called as Mesh network topology.

2. Tree network topology : The bus networks were joined together in a hierarchical manner like a tree structure. Therefore this topology was called Tree network topology.

A comparative study among Bus, Ring and Star Topology Network

Bus	Ring	Star
1. Each node is connected to single main cable.	1. Each node is physically connected to only two adjacent node	1. Each node is connected to centrally located server.
2. A fault in one of node does not affect the rest of network.	2. If the connection between two node fails entire network fails.	2. If server fails the entire network breaks dowm.
3. If fault in any node difficult to diagnose.	3. Easy to diagnose.	3. Very easy to diagnose.
4. Implementation cost is cheaper.	4. Moderate	4. Costilier.
5. More data traffic due to single cable dedication.	5. More data traffic.	5. Minimum data traffic.
6. Addition of few nodes affect time delay.	6. Addition of node affect time delay.	6. Addition of node does not affect time delay.
7. Direct communication between two node without any control station.	7. delay in communicating from one station to other station.	7. All communication between noe are through central node i.e. server.

Client - Server Networks :

The networks in which, the power is distributed and is inter connected by various devices are called as client-server networks. In this network the end user is called the client and the network is called server.

Inter Organisational Networks :

the networks, which connects the company to other companies or oranisation is called Inter organisational networks. Ex. for this type of networks are intranet, *extranet, internet.*

Internet

Internet can be defined as network of networks. It is a network, which connects the computer to the computers across the countries, which forms a big international network. thus one can retrive/share information from your desktop to outside world.

Information of Regarding Internet

Internet provides the fastest and cheapest medium for communication. One can store document, video and audio etc. It can also be used to carry out business activities and communicating with the companies of other countries.

In order to connect on the Internet you have to type up the path name for that network or particular website address. This path name is a very long one, which consistts of 32-bit number called IP addresses. Therefore this pathnames are given some easy names which relate them to the type of job they perform called Domain name. ou can view can look at the audio, video, and news or can carry out business and other activities through your computer by connecting to the specific site i.e. by specifying URL (Uniform Resource Locators) URL's are case sensitive.

URL is generally consist of four parts : (1) Protocal 92) Server or Domain name (3) Path and (4) File name

Internet Tools

Communication tools via E-mail, telnet, Usenet.

Information search tools via, Veronica, Archie, search engine etc.

Multimediua information tools via. World Wide Web.

Information retrieval tools via ftp and Gopher.

Types of Internet Connection :

E-mail connectivity

Dial-up Shell Account

Dial-up TCP/IP Account

Leased line

Requirements for Internet Connectivity :

Computer

Modem

Software called Web browser (Like, Internet Explore, Netscape Navigator, Mosaic etc.)

Phone Line

Internet service provider

Working of Internet :

The working of Internet depends on following

1. Cabling : Internet works, because it is connected to all the computers by some medium. This medium seems to be a large cable. Medium for communication of Internet now-a-days used are Telephone calbes or dedicated leased lines or microwave link.

2. languages : For transferring of information from one computer to another computers Internet uses several langaugea known as protocols. The protocols used by the Internet are Transmission Control Protocol (TCP) and Internet Protocol (IP) IP is responsible for detecting, when a messages are lost or damaged and getting it retransmitted. Thus TCP/IP decides the route and transmit the data in teh form of packets.

3. Software application : It is a common interface to the user with Internet. The page that is desplay on the screen on connecting to Internet is called as Web Brower. The popular Web Browsers are Netscape, Internet explorer.

Activities on Internet

Some of the activities that user can perform on Internet are listed below,

1. Sending and receiving mail messages.
2. Search the net for accessing inforamtion and down load them on PC, if required.
3. On-line chatting and joining contest.
4. On line shopping
5. On-line booking for various reservations (movie/airlines/railways)
6. Post your bio-data on nternet.
7. Playing on-line games
8. Creating web-sites.
9. Read and post articles for news groups.
10. Video/Audio conferencing etc.

E-commerce

E-commerce or E-com is the one of the most important tool on the internet for the business purposes. it is useful for the best secured business with great satisfaction of both the customer and the seller.

Definition

E-commerce is the group of tools and implementation techniques of IT used with some business strategies and pratice, so that business could be carried out electronically from anywhere in the best secured and easy way. Business over the net is called - e-commerce.

E-commerce support business from anywhere around the worldat any time. With these Internet tool even the smaller companies could compete with the major companies as it is cheaper and provides good satisfaction to the market.

Usefulness of E-commerce

It revolutioneses the way of business carried out earlier.

Other than buying and selling of items it also supports sharing collaborations with other organizations. It also supports activities like marketing, providing customer service, facilitating communications etc. The purchase could looks at the items displayed by various companies on the Internet.

Various features, all terms and conditions along with the cost and payment conditions of that specific item could be evaluated by the purchaser easily on Internet.

Without spending healthy expenditures for on-sight observations, the purchaser could meet their needs by sitting on net at their residence.

Other than these, all the meetings, collaborations and communications could directly be carried out on e-commerce.

The business carried out through e-commerce is best in all respects than otherwise.

Business with e-commerce saves time along with money.

Business with e-commerce is more versatile, productive and is open for 24 hours.

Application of E-commerce :

Business-to-Consumer (B2C) : These applications make business to come closer to the consumers or customers and vice-versa.

Business-to-Business (B2B) : These application involve electronic transactions for trade and business activities for businessmen.

Business-to-Government (B2G) : These applications involves business transactions with governmental agencies.

Consumer-to-Administration (C2A) : These applications involve inquiries of the customers with the administrative agencies without visiting them.

Difference between Traditional commerce and E-commerce.

Traditional Commerce	E-Commerce
1. Lot of paper work during business, fund transfer, purchase and selling of the products.	1. Paperless business transaction.
2. Consumer time and reduces efficiency during business transaction.	2. Reduce time, increase productivity and selling
3. More cost of products and services	3. It reduce the cost of product and services.
4. Scope of business is local	4. Scope of business is global.
5. Generally Service available between 9 AM to 9 PM in week days.	5. Service available on all days i.e. 24 x 7.
6. Knowledge of computer and Internet is not necessary	6. Knowledge of computer and Internet is must.
7. Deliveries of product is immediate due to local market	7. Late deliveries.

Characteristics of Computer in the Field of Research

To-day a researcher being an human being have weak teeth, feeble claws, no horns, cannot fly and cannot even run fast but nature has compensated all these shortcomings by giving us computer. It is because of computer that a researcher can calculate his results very fast. Computer is not only an extension of man's ability to compute but also help him in storing and retaining data for further research. Computer can even take logical decision on various aspects of research. In fact, the computer has come up in such a big way that its presence is felt in every shpere of research like research in education, research in business, research in trade, research in industry, research in medicine, research in banking, research in different services, etc, Now, we briefly examine the following capabilities of computer in the field of research.

1. Speed and Accuracy

A computer is a high speed device, capable of taking logical decisions, performing arithmetic and non-arithmetic operations for research. The high speed may be visualised by the fact that modern computer can perform billions of calculations relating to research in one second and the results of research are cent per cent accurate. It is perhaps the speed of the computer which helps a researcher to calculate fast. The reason for the extremely fast speed of a computer that it is an electronic device and works on electrical signals known as electric pulses.

An electrical pulse travels at extremely fast speed which in turn results into the fast speed of the computer. The high speed of computer can be judged from the fact that the work of indexing *The Encyclopedia Britannica* consisting of 30 volumes has taken 50 research scholars about 40 years to accomplish the job, but with the help of computer the job could be finished in less than one year. It is the high speed of computers which has made is possible for the researcher to land on the moon. Not only this, computer helps in research relating to whether forecasting as weather changes unpredictably, necessitaing very fast calculations and analysis. The results obtained by the computer relating to research are highly reliable and accurate.

2. Proceesing Capabilities

The computer is not only a fast calculator which outputs accurate results of research but it has the following characteristics :

(a) Capability to take logical decisions for research.

(b) Large and perfect memory relating to research data.

(c) Versatility

(d) Automation

(e) Diligence

(a) Capability to take Logical Decisions for Research

A computer has the capability to take decision relating to research, which imparts it some intelligence. A computer is capable of comparing data of research project, and depending upon the results of this comparision, it can take appropriate actions. It must be borne in mind that research data to be compared need not necessarily be numercial, a computer can handle non-numerical research data with the some ease at it handled the numerical data.

(b) Large and Perfect Memory Relating to Research Data

The most important characteristic of modern computer in which it scores over human brain is, its very large and perfect memory relating to research work. It is true that our brain can also store information correctly but its capacity is limited. On the other hand a computer is capable of storing a huge amount of research data and retrieving in instantly and that too with a perfect accuracy, A computer is capable of recalling the information stored in its memory at the rate of more than a million words in a fraction of a second.

(c) Versatility

A modern computer is a very versatile machine. Computer can be used to solve the prob;ems relating to various different research projects. At one instant, a computer may be solving a complete scientific research problem, the very next moment it may be playing a chess game and then be busy in taking up a pay-roll problem. In fact, a computer machine a capable of doing any task which can be split into a number of logical steps.

(d) Automation

A computer machine is automatic in operation. By saying that computer is automatic, we mean that once the research data and research instructions are fed to a computer, human interventions are not required. The computer manipulates the data according to the instruction of researcher and continues doing so automatically till the last instruction of researcher is executed.

(e) Diligence

A computer is a very hard working instrument. It has super high diligence and never feels tired or fatigue even if it has to work for very long hours. Thus, it may be said that computer is superio to human brain in this respect, because human brain feels tired after long working hours and some times finds it difficult to work further but a modern computer machine does not suffer from this handicap. Moreover even after working for long hours, there is no loss of accuarcy in its results which is a marvellous feature of computer.

The above charcteristics of computer machine make it an excellent device for research work.

Role of Computers in Research Process

The computers are indispensable throughout the research process. The role of computer becomes more important when the research is on a large sample. Data can be stored in computers for immediate use or can be stored in auxiliary memories like floppy disks, compact disks, pen drives or memory cards, so that the same can be retrieved later. The computer assists the researchers throughout different phases of research process.

Phase of Research Process

There are 5 major phases of the research process. They are–

(1) Conceptual phase.

(2) Design and Planning phase.

(3) Empirical phase.

(4) Analytical phase.

(5) Dissemination phase.

(1) Role of Computers in conceptual phase

The conceptual phase consists of formulation of research problem, review of literature, theoretical framework and formulation of hypothesis.

Role of computers in literature review/computers help in searching the literatures (for review of literatures) and bibliographic references stored in the electronic databases of the World Wide Web. It can thus be used for storing relevant published articles to be retrieved whenever needed. This has the advantage over searching the literatures in the form of books, journals and other news letters at the libraries which consumes considerable amount of time and effort.

(2) Role of Computers in designing and planning phase:

Designing and planing phase consists of research design, population, research variables, sampling plans, reviewing research plans and pilot study.

Role of computers for sample size calculations: Several softwares are available to calculate the sample size required for a proposed study. NCSS-PASS-GESS is such software. The standard deviation of the data from the pilot study is required for the sample size calculation.

(3) Role of computers in Empirical Phase

Empirical Phase consist of collecting and preparing the data for analysis.

Data Storage: In research studies, the preparation and inputting data is the most labor intensive and time consuming aspect of the work. Typically the data will be initially recorded on a questionnaire or record form suitable for its acceptance by the computer. To do this, the researcher, in conjunction with the statistician and the programmer will convert the data into Microsoft Word file or Excel Spreadsheet. These spreadsheets can be directly opened with statistical software's for analysis.

(4) Role of Computers in Data Analysis:

This phase consists of statistical analysis of the data and interpretation of results.

Data analysis: Many softwares are now available to perform the mathematical part of the research process i.e. the calculations using various statistical methods. Softwares like SPSS, NCSS-PASS, STATA and Sys at are some of the widely used softwares.

They can be used in places like calculating the sample size for a proposed study, hypothesis testing and calculating the power of the study. Familiarity with any one package will suffice to carry out the most intricate statistical analysis.

Computers are useful not only for statistical analysis but also to monitor the accuracy and completeness of the data as they are collected.

(5) Role of computers in dissemination:

This phase is the publication of the research study.

Research Publishing: The research article is typed in word format and converted to portable data format (PDF) and stored or published in the World Wide Web.

To conclude, computers are useful tools that make the research process easier and faster with accuracy and greater reliability and fewer errors. The programmer or the computer operator should have a thorough knowledge about the abilities and limitation of the software used for better use of computers.

QUESTIONS

1. What is computer? Explain the difference between digital computer and Analog Computer.
2. Differentiate between hardware and software?
3. Explain the utility of Binary numbers as used in computers.
4. Differentiate between Traditional commerce and E-commerce
5. Discribe some of the important applications and uses of computers in present times.
6. Explain the capabilities of computer in the field of research.

■■■